Contents

£38·50.
£ ML

Convergent Beam Electron Diffraction of
Alloy Phases

Convergent Beam Electron Diffraction of Alloy Phases

by

The Bristol Group
under the direction of
John Steeds

Compiled by John Mansfield

This atlas is a compilation of work by the following members of the University of Bristol, Department of Physics, Electron Microscopy Group:

Iain Brock   Mervyn Shannon
Kevin Cooke   John Steeds
Alwyn Eades   Larry Stoter
Nigel Evans   Roger Vincent
Yang-Pi Lin   John Walmsley
John Mansfield

Adam Hilger Ltd, Bristol and Boston

© Adam Hilger Limited 1984

British Library Cataloguing in Publication Data

Electron Microscopy Group
  Convergent beam electron diffraction of alloy phases.
  1. Physical metallurgy 2. Alloys——Analysis 3. Electrons——Diffraction 4. Phase rule and equilibrium
  I. Title   II. Steeds, John   III. Mansfield, John
  669'.94   TN690

ISBN 0-85274-771-3

Consultant editor: **Dr G Nickless**, School of Chemistry, University of Bristol

Published by Adam Hilger Ltd
Techno House, Redcliffe Way, Bristol BS1 6NX, England
PO Box 230, Accord, MA 02018, USA

Typeset and printed in Great Britain by The Alden Press, Oxford

Preface

One of the most useful functions of transmission electron microscopy is the identification of the phases present in a sample. This is particularly important in the case of small particles in a multi-phase material where other techniques are not available.

It is now possible to make quick and routine identification of phases without elaborate analysis or calculation. Two techniques, whose widespread application is relatively new, have brought this about. These techniques are convergent beam diffraction and energy dispersive x-ray analysis. Although both have limitations (convergent beam diffraction can identify only crystalline material of reasonable perfection and energy dispersive x-ray spectroscopy detects only elements heavier than neon) both techniques have a most important virtue: they are 'fingerprint' techniques. Once appropriate standards have been established, both convergent beam patterns and energy dispersive spectra can be readily recognised, to give immediate identification of the phase under observation. Each zone axis of each crystal structure produces a zone axis pattern that is distinct and *recognisably* distinct (this overstates the situation but not by much), and in many cases the x-ray spectrum is also sufficiently characteristic of a phase to identify it without further analysis.

The drawback, if any, lies in the qualifier: once appropriate standards have been obtained. When a phase is observed for the first time, its identification is more complex and may, in some cases, require the synthesis of a sample of the phase for the purpose of comparison. This is the hard part of the process and it is in the hope of minimising repetition of work which we have already performed that this atlas has been prepared. Over a period of several years, members of the Electron Microscopy Group at the University of Bristol have worked on a number of alloys, particularly stainless steels and superalloys. As a result, a set of standards has been generated for the range of phases that occur in these alloys. These standards have been gathered together in a standard format and form the body of the book. It is hoped that workers in other laboratories who complete standard sets of convergent beam patterns and energy dispersive x-ray spectra will have them published in a format as close to that of the present atlas as the journal will permit. In this way, reprints of the articles can be added to this book and the number of phases with their fingerprints on file will be extended.

Acknowledgments

The production of this atlas would not have been possible without the support and encouragement of a number of organisations. We would first like to express our thanks to the Science and Engineering Research Council for financial support during the actual correlation and compilation process and for the original research studentships in which much of the data was recorded.

Most of the work contained in the text has been stimulated by collaborative projects undertaken with Berkeley Nuclear Laboratories, AERE Harwell, Rolls Royce, Exxon Corporation and Rockwell International. We would like to thank all those concerned for their assistance, financial support and the provision of specimens.

dimensional diffraction are associated with very long extinction lengths and do not change much with thickness for thicker crystals in the range 80–300 nm.

The material on each of the phases referred to in this text has been set out in a standard fashion. Crystallographic data, composition and morphology of each phase is given together with any special features which were noted. Energy dispersive x-ray spectra, a convergent-beam electron diffraction 'zap map' and chosen diffraction details are included as aids to identification. The order of presentation is matrix, intermetallics, carbides, nitrides, borides, sulphides, phosphides, oxides and slags. The conventions used for Miller and Miller–Bravais indices are set out in appendix 3 and in appendix 4 various crystallographic formulae are given. Appendix 1 contains a summary of abbreviations used.

4 Energy dispersive x-ray (EDX) spectra

The EDX spectra presented here were all obtained from particles extracted on carbon replicas, except for matrix spectra, the δ ferrite spectrum and one of the $M_{23}X_6$ spectra, which were obtained from thin electropolished foils. The quality of spectra from extracted particles greatly exceeds that which can be obtained from analyses of particles embedded in the parent matrix, and this fact should be kept in mind when using the data. As an example, note the spectrum for Cr_2N (figure 12.1), which contains essentially no spurious peaks.

The reasons for the superiority of spectra obtained from extracted particles is quite well understood. We were working with a detector at zero take-off angle and had therefore to tilt the specimens towards the detector in order to obtain x-ray counts. Back-scattered high-energy electrons generated at the region of interest spiral about the optical axis in the lens magnetic field and strike the specimen and supporting copper grid at points remote from the selected region. In the case of well separated, thin particles on a carbon extraction replica the integrated effect of the spurious signals generated in this way is negligible. However, for electropolished discs, which are often used, it is likely that most of the spiralling, high-energy electrons will strike thick parts of the disc, producing high x-ray generation rates and serious degradation of the spectra.

It is for similar reasons that the grid material appears so strongly in the spectra. Even though the total grid area covered by grid material is not very great, that material is so thick as to produce strong x-ray emission. Many of the spectra presented were generated with a beryllium holder permitting a small tilt of approximately 20° for reasonable x-ray count rates. Other holders were also used; these required greater tilts and gave poorer spectra both on

account of the tilt angle and because of x-ray generation from the material of the holder adjacent to the specimen. A few results were obtained with extraction replicas supported on beryllium grids. We have made no attempt to remove the spurious peaks evident in most of our spectra. Copper and zinc peaks on the spectra are regarded as spurious and are unlabelled in the material presented.

It is also established that the EDX spectra depend on specimen thickness [13], at least for thickness above a certain minimum. While aware of this point, we have not made a systematic study of the dependence of any of our EDX spectra on specimen thickness. We believe that the spectra presented here were obtained from sufficiently thin samples that absorption effects were not important. We also note that EDX spectra are sensitive to crystallographic orientation. Several workers have published results that clearly demonstrate that there can be a large difference in the emitted x-ray intensities, depending upon the diffracting condition of the specimen [14–16]. This intensity variation is related to the relative excitation of Bloch waves close to atomic cores. For a detailed description, the reader is directed to references [14–16]. It will suffice to comment here that the spectra which we have included were recorded with the specimens tilted away from strong diffracting conditions; any attempt that the reader makes to analyse phases by using EDX 'fingerprints' should involve similar precautions.

We have not given much attention to very weak peaks in the spectra, those that are only just above the noise level. Weak vanadium peaks have been found quite frequently in certain phases, such as M_6X and M_3B_2, although vanadium levels were considered to be negligible in the alloys concerned. There is, at the time of going to press, a recent development concerning weak x-ray peaks which deserves comment. The bremsstrahlung radiation background has traditionally been thought of as a smooth function of energy. However, it has been known for a long time [17] that when a high-energy charged particle was incident along low index direction in a crystal the bremsstrahlung curve developed peaks at specific energies. These peaks are known as coherent bremsstrahlung. The same effect has now been observed at 100 kV in EDX spectra [18]. It is easy to establish whether a weak peak of interest results from the coherent bremsstrahlung process. If the microscope operating voltage were changed, the peak of interest would shift in energy by an amount proportional to the change of electron velocity. Alternatively, if the sample were tilted through about 5° and a new spectrum acquired, the coherent bremsstrahlung, sensitive to direction of incidence, would either vanish or at least shift to a new and very different energy.

Where a phase was found to have a wide range of

Introduction

1 Convergent beam electron diffraction—what is it?

A convergent beam pattern is a transmission electron diffraction pattern obtained by observing or photographing the diffraction plane (the back focal plane) in a transmission electron microscope. In this it is like any other form of electron diffraction in an electron microscope (e.g. selected area diffraction); it differs from the other methods in that the specimen is illuminated with a convergent beam of electrons instead of a parallel beam. The area from which the diffraction is obtained is determined by the size of the small probe formed by focusing the incident illumination in the plane of the specimen. We assume incoherent illumination conditions here and throughout this introduction, since these result from using heated tungsten filaments or LaB_6 sources. Typical probe sizes for generating the patterns illustrated in this text were in the range 20–50 nm.

The resulting diffraction pattern contains an array of discs. All the discs are the same size and shape (ignoring any distortion produced by the microscope) and there is a disc associated with each Bragg diffracted beam produced by the crystal. Each disc is a map of the variation of diffracted intensity with angle of incidence. Each angle of incidence corresponds to a particular point within a given disc and a set of equivalent points in each disc of a pattern. The diameter of the discs is proportional to the angle of convergence of the electrons incident on the specimen and this is controlled by the diameter of the second condenser aperture (see, for example, figures 6.3 and 6.4). Between the array of discs there is a variable amount of diffuse scattering (see, for example, figures 1.3 or 3.3(b)) resulting from thermal or static disorder of the crystal (phonon or defect scattering). For thick crystals the diffuse scattering may be so strong as to obscure completely the perimeter of the discs (images of the second condenser aperture). An example of this can be found in figure 19.2. The really sensitive information conveyed by convergent beam diffraction is contained within the discs themselves or as intensity differences between discs, and it is that which we shall be concentrating on here. An upper limit to the

degree of convergence obtainable is generally set by the need to prevent the different orders of reflection from overlapping, although it can also derive from an instrumental limitation. A large angular range is desirable in order to study the detailed intensity distribution within a given order. Techniques have recently been devised for overcoming this angular limitation [1, 2] but, as they have considerably poorer spatial resolution, they are not suitable for the study of small particles and we have not included any results here.

We have found that crystallographic analysis is most easily and reliably performed by orienting the particle of interest so that the electron beam is incident along a low index direction (see appendix 2 for the conventions adopted here for referring to such directions). The patterns used are properly referred to as zone axis patterns and not pole patterns. A zone axis is a line of common intersection of a number of distinct crystallographic planes; this is the same as saying that it is the real space direction perpendicular to a cross grating diffraction pattern. The reciprocal lattice vectors of which the cross grating pattern is composed are the normals to diffracting planes intersecting along the zone axis. However, a pole is the normal to a crystallographic plane. In the case of cubic crystals the $\langle uvw \rangle$ zone axis and pole are identical but in crystals of lower symmetry they are generally distinct. For this reason it is strongly recommended that the term pole be dropped in favour of zone axis in referring to cross grating diffraction patterns.

Most electron microscopists are familiar with selected area diffraction. The question often arises as to the ways in which convergent beam electron diffraction is a superior technique. In fact, for measuring angles between reflections, identifying weak satellite peaks, or studying diffuse scattering, there is either no advantage or else a distinct disadvantage to using convergent beam diffraction. However, the selected area technique generates patterns where the diffracted beams have been averaged over thickness and angle in a complicated way which

1

composition, several EDX spectra are included to illustrate typical results. Our detector was fitted with a thin protective beryllium window so that all characteristic peaks from light elements were attenuated away. The performance of the detector was investigated by a droplet technique [19] and the resulting sensitivity curve is shown in appendix 5.

In general, once a phase has been unambiguously identified by electron diffraction its EDX fingerprint will usually be found the most convenient means of subsequent identification. Clearly the lack of light element information is something of a handicap in this respect. An isolated chromium peak might come from the metal, an oxide, nitride or carbide, for example. Moreover, some complex phases have variable composition ranges which extend to regions of overlap. An example of this is provided by $M_{23}X_6$ and M_2B which can have the same metallic component. One rather distinctive element is silicon. Apart from slags and SiC, silicon has been found only in M_6X, Laves phase, $M_{23}X_6$ (occasionally) and AlN (small component). High molybdenum concentrations have been noted in M_3B_2, Laves phase and M_6X and quite high concentrations were found in σ phase and (TiMo)C.

References

1 Tanaka M, Saito R, Ueno K and Harada Y 1980 *J. Electron Microsc.* **29** 408

2 Eades J A 1980 *Ultramicroscopy* **5** 71

3 Buxton B F, Eades J A, Steeds J W and Rackham G M 1976 *Phil. Trans. R. Soc.* **281** 171

4 Steeds J W and Vincent R 1983 *J. Appl. Crystallogr.* **16** 317

5 Eades J A, Shannon M D and Buxton B F 1983 Crystal Symmetry from Electron Diffraction in *Scanning Electron Microscopy* (Chicago: SEM Inc, A M F O'Hare)

6 Tanaka M, Saito R and Sekii H 1983 *Acta Crystallogr.* A **39** 357

7 Shannon M D and Steeds J W 1977 *Phil. Mag.* **36** 279

8 Steeds J W 1984 'Electron Crystallography', lectures delivered at the 25th Scottish Universities Summer School in Physics, to be published

9 Steeds J W and Vincent R 1983 *J. Microscopie Spectrosc. Electron.* **8** 419

10 Gjønnes J and Moodie A F 1965 *Acta Crystallogr.* **19** 65

11 Steeds J W, Rackham G M and Shannon M D 1978 *Electron Diffraction 1927–1977* (Inst. Phys. Conf. Ser. 41) p 135

12 Tanaka M, Sekii H and Nagasawa T 1983 *Acta Crystallogr.* A **39** 825

13 Goldstein J I 1979 *Introduction to Analytical Electron Microscopy* ed J J Hren, J I Goldstein and D C Joy (New York: Plenum) ch 3

14 Hall C R 1966 *Proc. R. Soc.* A **295** 140

15 Cherns D, Howie A and Jacobs M H 1973 *Z. Naturforsch.* **28a** 565

16 Bourdillon A J, Self P G and Stobbs W M 1980 *Electron Microscopy and Analysis 1979* (Inst. Phys. Conf. Ser. 52) p 429

17 Ter-Mikaelian 1971 *High Energy Electromagnetic Processes in Condensed Media* (New York: Wiley)

18 Spence J C H, Reese G, Yamamoto N and Kuriski G 1983 *Phil. Mag.* B **48** L39

19 Graham R and Steeds J W 1984 *J. Microsc.* **133** 275

Bibliography

Eades J A 1980 *Physics of Modern Materials* (Vienna: Int. At. Energy Agency) vol I, pp 31–40

Goodman P 1980 *Scanning Electron Microscopy 1980* (Chicago: SEM Inc, A M F O'Hare) vol I, p 53

Kossel W and Möllenstedt 1939 *Ann. Phys., Lpz* **36** 113

Lempfuhl G 1978 *Electron Microscopy 1978* (Microsc. Soc. Can.) vol III p 304

Steeds J W 1979 *Introduction to Analytical Electron Microscopy* ed J J Hren, J I Goldstein and D C Joy (New York: Plenum) ch 15

Steeds J W 1984 'Electron Crystallography', lectures delivered at the 25th Scottish Universities Summer School in Physics, to be published

Stoter L P 1981 *J. Mater. Sci.* **16** 1356

The Atlas

Austenite or γ

Phase name: austenite or γ
Lattice type: fcc
Point group: *m3m*
Space group: *Fm3m*
Lattice parameters: $a = 0.3585$–0.3598 nm in 316 steel [1,2,3]
$a = 0.3580$–0.3620 nm in super-alloys [4]

FORMULA AND COMPOSITIONAL VARIABILITY

Austenite exists over a wide range of compositions. For details of phase diagrams and general references see [2] and [5]. The specifications for the alloys that have been most frequently studied in the work summarised here are listed in table 2 of appendix 2.

OCCURRENCE

It exists as the matrix phase for many high alloy stainless steels and superalloys.

REFERENCES

1 Weiss B and Stickler R *Met. Trans.* **3A** 851
2 Peckner D and Bernstein I M (ed) 1977 *Handbook of Stainless Steels* (New York: McGraw–Hill), particularly chapter 4
3 Andrews K W, Dyson D J and Keown S R 1971 *The Interpretation of Electron Diffraction Patterns* (London: Adam Hilger)
4 Kriege O H and Baris J M 1969 *Trans. ASM* **62** 195
5 Sims C T and Hagel W C (ed) 1972 *The Superalloys: Vital High Temperature Gas Turbine Materials for Aerospace and Industrial Power* (New York: Wiley)
6 Shannon M D and Steeds J W 1971 *Phil. Mag.* **36** 279
7 Sarikaya M, Thomas G, Steeds J W, Barnard S J and Smith D W 1981 *Solute Partitioning and Austenite Stabilisation in Steels: Int. Conf. on Phase Transformations, Pittsburgh, USA* (Ohio: American Society for Metals)

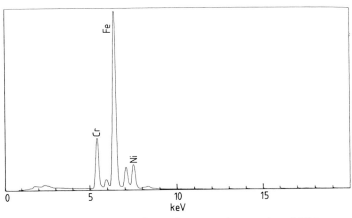

Figure 1.1 EDX spectrum from the austenite matrix of 316 stainless steel.

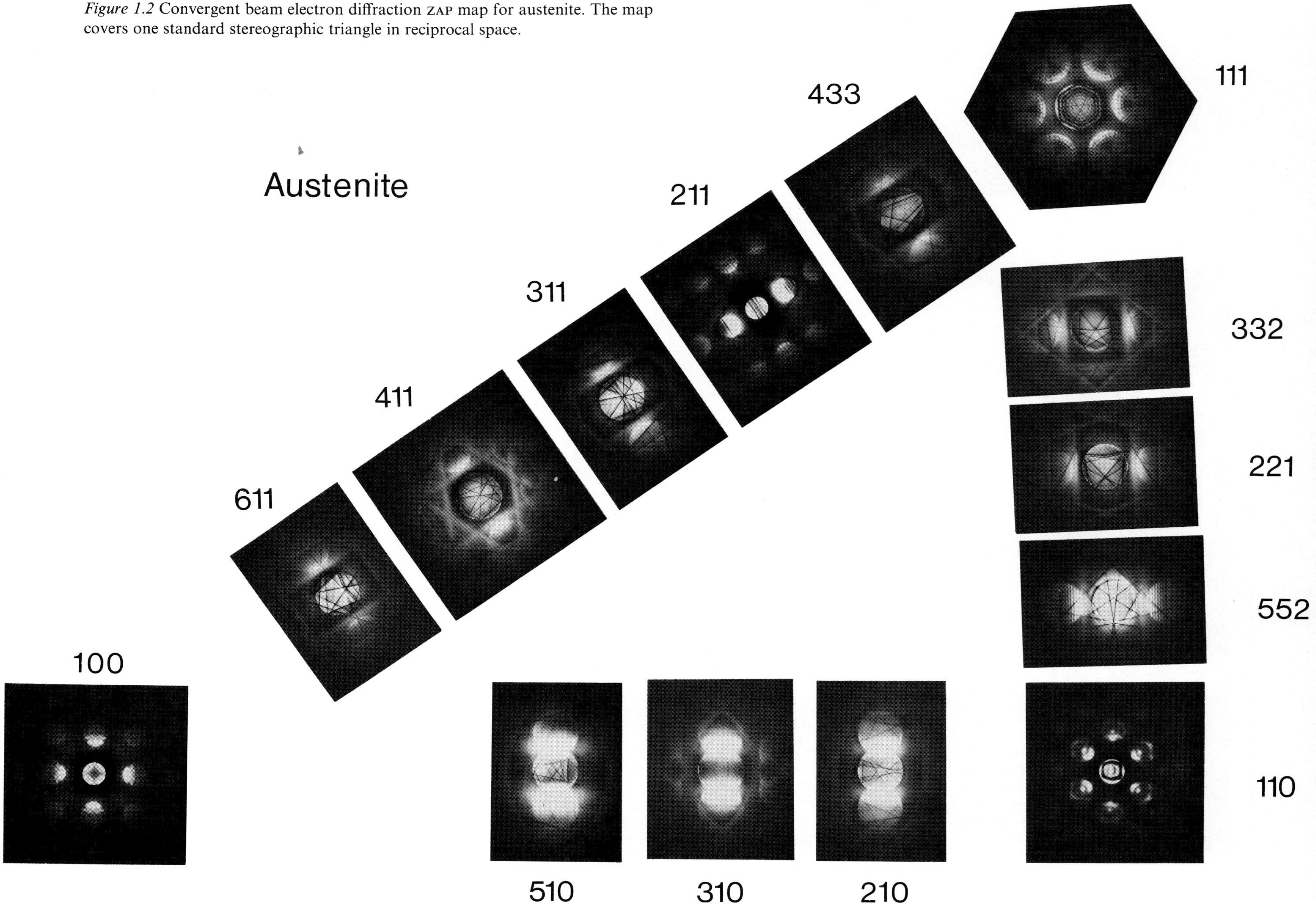

Figure 1.2 Convergent beam electron diffraction ZAP map for austenite. The map covers one standard stereographic triangle in reciprocal space.

20

40

60

80

Figure 1.3 Austenite ⟨111⟩ thickness sequence, taken from a specimen with a shallow wedge angle. As the probe was brought in from the edge there was a steady increase in specimen thickness and associated expansion of the system of concentric rings in the bright field disc. A new ring appears at the centre of the pattern for every increase in thickness of one extinction distance [6]. For this axis the extinction distance length has been calculated as 20 nm. The micrographs are labelled with approximate thicknesses using this value. Note that one cannot just count the number of rings to determine the specimen thickness

100

120

140

160

because as the thickness increases the outermost ring periodically spills off the edge of the pattern.

The HOLZ lines in the central disc of this pattern have been used by Sarikaya *et al* [7] to study the carbon content of retained austenite in a martensitic steel. Since there is no absolute value for the lattice parameter of austenite, they used pure Ni and Cu as standards.

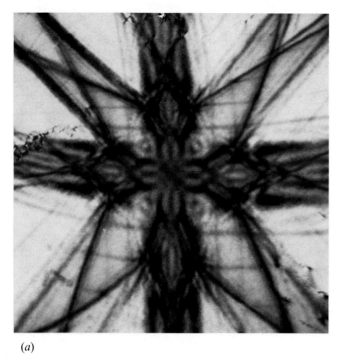

(a)

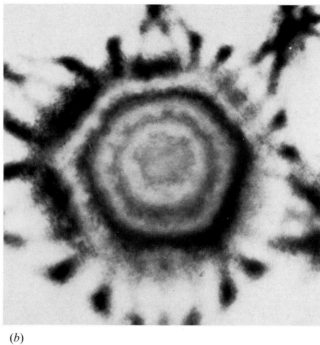

(b)

(c)

Figure 1.4 A selection of bend-contour ZAPs from the matrix of 316 steel. Although the symmetry of these patterns can be determined, it is often difficult when the material is irregularly buckled. HOLZ lines are not often seen at the centre of low index zone axes, principally because of the convergence of the illumination. (a) $\langle 100 \rangle$. (b) $\langle 111 \rangle$. (c) $\langle 110 \rangle$. It is interesting to compare these patterns with those from ferrite (§2). Many differences and just a few similarities can be observed.

(a)

(b)

(c)

(d)

Figure 1.5 Long camera length CBD ZAPs for the major axes of austenite. (*a*) ⟨100⟩. The symmetry here, which should be 4*mm*, is not perfect because the pattern is not ideally centred. (*b*) ⟨110⟩. The centre of this pattern is distorted due to buckling in the foil from which it was recorded. (*c*) ⟨211⟩. (*d*) ⟨411⟩.

13

Diamond

Phase name: diamond
Lattice type: diamond cubic
Point group: $m3m$
Space group: $Fd3m$
Lattice parameter: $a = 0.3575$ nm [1]

OCCURRENCE AND MORPHOLOGY

Diamond was occasionally found as an artefact on CERs produced from steel specimens which had been polished with diamond paste. It is included here because it has a similar lattice parameter to austenite. Although austenite is not frequently observed on replicas, it is possible that the two phases could be confused if studied by conventional SAD. Examination of the CBD patterns, however, shows that the two phases are readily distinguishable.

REFERENCE

1 Donnay J D H and Ondik H N (ed) 1973 *Crystal Data Determinative Tables, Vol 2—Inorganic Compounds* (US Dep. Commer., NBS and JCPDS)

(a)

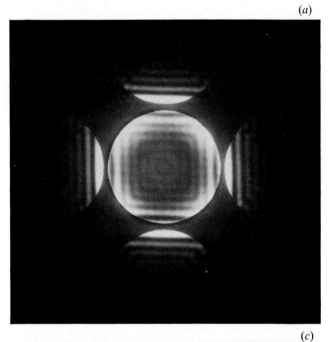

Figure 1.6 Three major diamond zone axis patterns. (a) ⟨100⟩. The fringes in this pattern are clearly different from those in the austenite ⟨100⟩ axis (figure 1.5). Note that the reflections nearest to the central disc are of 220 type and not 200 type because 200 reflections are forbidden in diamond. (b) ⟨110⟩. The projected potential for this axis has a 'dumbbell' shape which results in a different pattern in the central disc as compared with austenite, which has a 'simple string' projection (see Introduction) along its ⟨110⟩ axis. (c) ⟨111⟩. At 100 kV it is easy to distinguish between the austenite and diamond ⟨111⟩ axes. In particular the HOLZ detail in the central disc is quite distinctive (compare this pattern with those in figure 1.3). At other voltages, e.g. 120 kV, the difference may be rather more subtle and care should be taken in identifying the patterns.

(b)

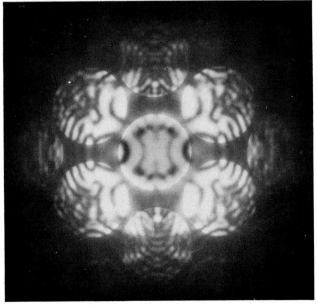

(c)

Phase name: ferrite, α or δ
Lattice type: bcc
Point group: $m3m$
Space group: $Im3m$
Lattice parameters: $a = 0.2866$ nm in steel, δ-ferrite [1]
$a = 0.2878$ nm in IN 718 superalloy, α-ferrite [2,3]

Ferrite α and δ

FORMULA AND COMPOSITIONAL VARIABILITY

Ferrite exists over a range of compositions, particularly in alloys with a high proportion of body-centred metals. For examples of phase diagrams see [4] and [5].

MORPHOLOGY

δ ferrite occurred in 316 steel as large irregular blocks and continuous grain boundary sheets. α ferrite was found in IN 718 as acicular plates, sometimes growing in close association with δ phase.

OTHER FEATURES AND COMMENTS

α ferrite has also been termed α'. The exact term depends upon the alloy system in question.

OCCURRENCE

δ ferrite was found in a type 316 steel that had been in service in steam generating plant for 28 000 hours at 650 °C. α ferrite was discovered in fully aged IN 718.

REFERENCES

1 Andrews K W, Dyson D J and Keown S R 1971 *The Interpretation of Electron Diffraction Patterns* (London: Adam Hilger)
2 Baker J F, Ross E W and Radavich J F 1970 *J. Metals* **22** 31
3 Raudebaugh R J and Sadowski E P 1959 *Trans. Met. Soc. AIME* **215** 23
4 Peckner D and Bernstein I M (ed) 1977 *Handbook of Stainless Steels* (New York: McGraw–Hill), particularly chapter 5
5 Sims C T and Hagel W C (ed) 1972 *The Superalloys: Vital High Temperature Gas Turbine Materials for Aerospace and Industrial Power* (New York: Wiley)

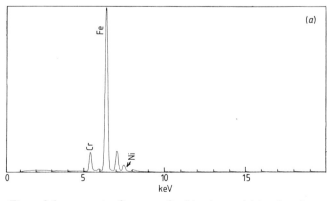
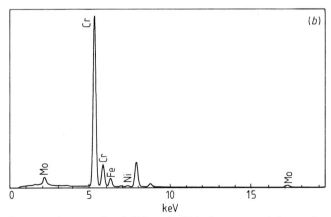

Figure 2.1 EDX spectra from two ferritic phases. (*a*) is taken from a lath of δ ferrite that had formed at an austenite boundary in a long-service sample of 316 steel. (*b*) is from a precipitate of α ferrite found in a sample of IN 718.

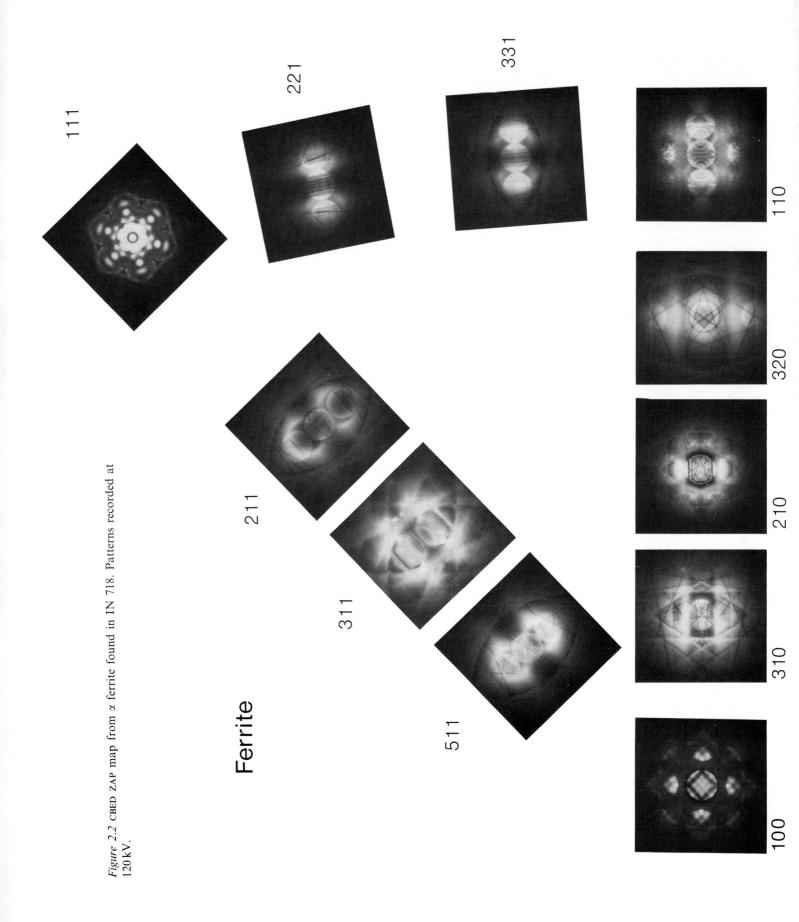

Figure 2.2 CBED ZAP map from α ferrite found in IN 718. Patterns recorded at 120 kV.

(a)

(b)

(c)

Figure 2.3 Extinction contour patterns recorded from δ ferrite discovered in a sample of 304 stainless steel. It is interesting to compare these with the equivalent patterns from γ (see figure 1.4). (*a*) $\langle 100 \rangle$. (*b*) $\langle 111 \rangle$. (*c*) $\langle 110 \rangle$.

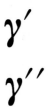

γ'

γ''

Phase name: The matrix strengthening phases in super-alloys are called γ' and γ''

Lattice type: γ': primitive cubic γ'': bct

Point group: γ': $m3m$ γ'': $4/mmm$

Space group: γ': $Pm3m$ γ'': $I4/mmm$

Lattice parameters: γ': $a = 0.358{-}0.362$ nm; note this range is the same as that quoted for the superalloy matrix (γ). This is because the difference between the lattice parameters for γ and γ' is rarely greater than 1% [1,2].

γ'': $a = 0.362$ nm, $c = 0.741$ nm, for IN 718. Note that c is approximately twice the cubic cell size [3].

FORMULA, COMPOSITIONAL VARIABILITY AND OCCURRENCE

The strengthening effect of these phases is dependent on the alloy composition as well as its heat treatment. Composition has a direct correlation with the lattice parameters of γ, γ' and/or γ'' phases. It is the degree of coherency or mismatch which determines the strengthening effect.

γ' is the most common strengthening phase in superalloys. In most nickel-based alloys it is based on $Ni_3(Al, Ti)$ with the Al/Ti ratio closely resembling that of the bulk. Ni in γ' may be replaced by Cr, Co, V, Mo, etc. In Ni–Fe based superalloys, such as Incoloy 901, γ' is generally based on Ni_3Ti.

γ'' is the major strengthening phase of the Ni–Fe based alloy Inconel 718 where it has the approximate composition Ni_3Nb.

MORPHOLOGY

γ' generally spherical or cuboidal in form.
γ'' generally occurs as discs or platelets.

OTHER COMMENTS AND FEATURES

During prolonged aging or exposure at service temperatures γ' may transform to the hexagonal phase that is known as 'η phase' [2]. This phase has the same space group ($P6_3/mmc$) as the Laves referred to in §6 but a different atomic arrangement. Whereas the common Laves phase has the $MgZn_2$ structure and is sometimes called η phase, the hexagonal phase formed by transformation of γ' has the arrangement of Ni_3Ti. It forms either intergranularly in cellular form or intragranularly as acicular plates. It has not been observed in the alloys we have studied.

Under similar conditions γ'' also transforms to the orthorhombic δ phase (§4) which has the same composition as the γ''.

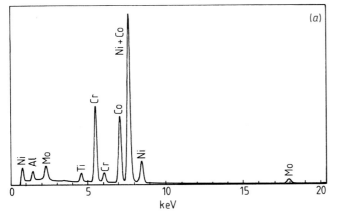

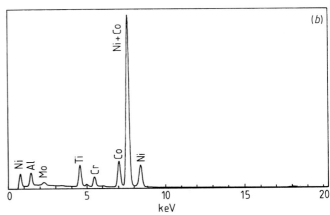

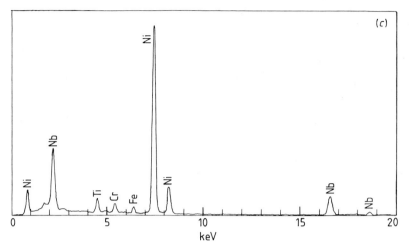

Figure 3.1 The three EDX spectra opposite and above are from: (*a*) γ' and γ together (in Astroloy). The γ is present as a thin layer between the γ' particles. Hence a probe much smaller than the 10 nm used here would have been required to obtain an EDX spectrum of the γ alone. (*b*) γ'. The γ' particles (again in Astroloy) were several μm in length so that a reliable spectrum could be obtained without difficulty. (*c*) γ'' particles found on a CER of an electron beam weld in IN 718. In IN 718 the mixture of γ and γ'' was on such a fine scale that it was impossible to exclude the possibility that there was some γ' also present. It was not possible to distinguish any of these phases by EDX analysis of a thin foil.

REFERENCES

1 Kriege O H and Baris J M 1969 *Trans. ASM* **62** 195
2 Sims C T and Hagel W C (ed) 1972 *The Superalloys: Vital High Temperature Gas Turbine Materials for Aerospace and Industrial Power* (New York: Wiley)
3 Paulonis D F, Oblak J M and Duvall D S 1969 *Trans. ASM* **62** 611
4 Vincent R and Bielicki T A, 1982 Applications of CBD to Nickel Superalloys in *Electron Microscopy and Analysis 1981* (Inst. Phys. Conf. Ser. 61) p 491

(*a*)

(*b*)

Figure 3.2 CBED patterns recorded from a thin foil of Astroloy. The patterns were taken at long camera length with the instrument operated in the STEM mode. This permitted the use of a probe small enough to select the narrow regions of γ that were not overlapped by other phases (these regions were less than 20 nm wide). (*a*) is from the γ' strengthening phase and (*b*) is from the matrix γ phase. The difference in lattice parameters between the two phases is quite small ($<1\%$) but it can just be detected in the bright field HOLZ deficiency lines present in the figures. The most obvious feature that distinguished these two phases was the presence of the 110 superlattice reflections in the γ' patterns (primitive cubic lattice). Patterns recorded at 120 kV.

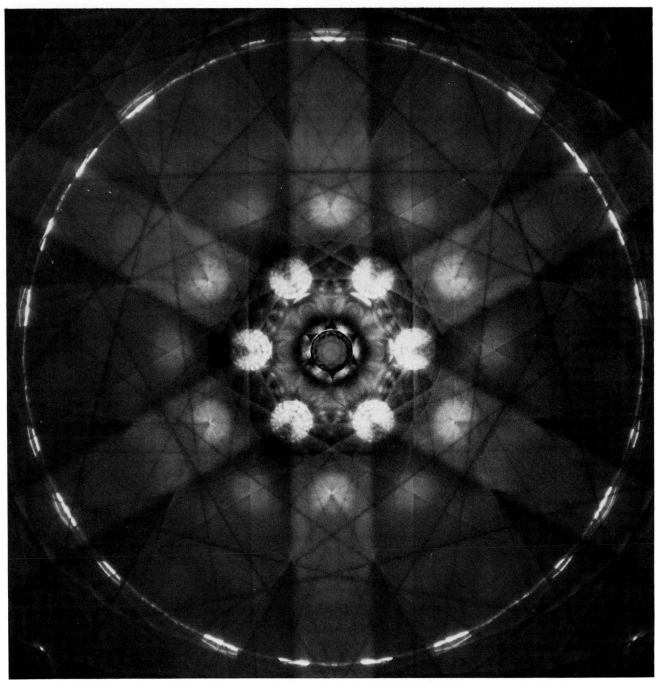

Figure 3.3(a) $\langle 111 \rangle$ axis of γ'. As the phase is primitive cubic weak 110 reflections are just visible mid-way between the centre of the pattern and the strong 220 reflections. The symmetry of the whole pattern is 3*m*.

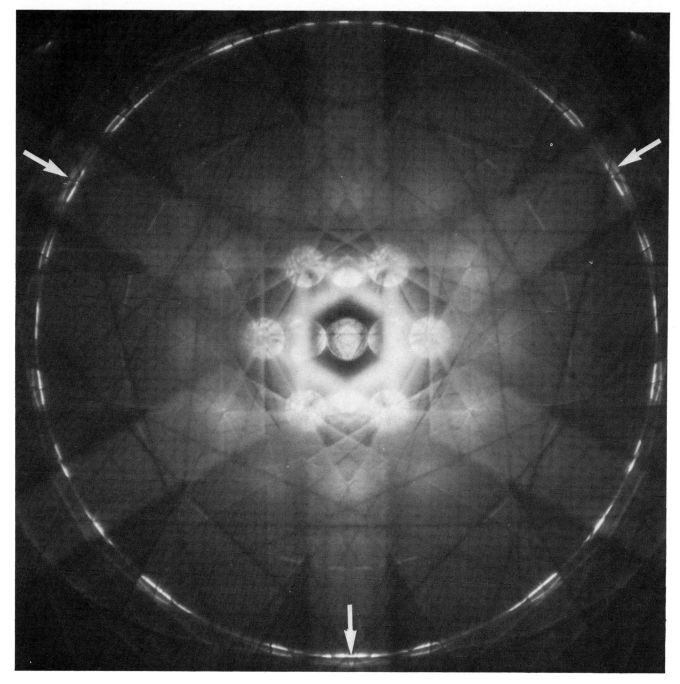

Figure 3.3(b) ⟨221⟩ axis of γ″. But for the ordering present in this phase, this would be the ⟨111⟩ axis of the cubic cell, i.e. the equivalent of (*a*). In this case the tetragonal symmetry dictates that only one set of the three distinct 220 rows of the disordered phase is divided by weak additional reflections. This divided row is indexed as hh̄0, the other two being h02̄h and 0h2̄h. Evidence

for the mirror symmetry of the pattern can be found in several places: for example, note the distortion in the Kikuchi intersections marked on the micrograph in white [4]. A characteristic feature of this pattern is the weak FOLZ ring of small radius which is a consequence of ordering in the tetragonal cell. Patterns recorded at 120 kV.

(a)

(b)

Figure 3.4 This figure illustrates two zone axes of γ'' which should be compared with the equivalent axes of γ (figure 1.5). (a) $\langle 201 \rangle \gamma''$. This pattern is equivalent to the $\gamma \langle 101 \rangle$ axis and is hard to distinguish from it (see figure 1.5(*b*) for comparison). (*b*) γ'' [001] axis. Weak flaws have been introduced during the development of the negatives. This pattern is similar to that from the $\gamma \langle 100 \rangle$ axis (figure 1.5(*b*)). 110 superlattice reflections centre the square mesh of fcc matrix reflections. Patterns recorded at 120 kV.

Phase name: δ phase (beware of confusion with δ ferrite in steels)

Lattice type: orthorhombic, Cu_3Ti structure [1]

Point group: *mmm*

Space group: *Pmmn*

Lattice parameters: $a = 0.5064$ nm, $b = 0.4224$ nm, $c = 0.4448$ nm in Ni_3Mo [2]
$a = 0.5136$ nm, $b = 0.4248$ nm, $c = 0.4530$ nm in Ni_3Nb [3]

δ

FORMULA AND COMPOSITIONAL VARIABILITY

The composition of this phase was dependent on the alloy matrix composition. In IN 718 it was basically Ni_3Nb, often with traces of Fe, Cr and Ti [4]. In the Ni–Al–Mo alloys studied by Kaufman *et al* [3] the composition was approximately Ni_3Mo, usually with traces of aluminium.

MORPHOLOGY

Globular precipitates are deliberately introduced by heat treatment to increase the ductility (at the expense of a slight decrease of strength). However, lath or plate-like intragranular δ phase precipitates can form during aging and they are detrimental to the mechanical properties.

OCCURRENCE

δ phase forms in superalloys either by transformation from γ'' during prolonged aging or directly by heat treatment at very high temperatures (compare with the formation of δ ferrite in austenitic steels) [5].

REFERENCES

1 Donnay J D H and Ondik H N (ed) 1973 *Crystal Data Determinative Tables, Vol 2—Inorganic Compounds* (US Dep. Commer., NBS and JCPDS)
2 Saito S and Beck P A 1959 *Trans. TMS-AIME* **215** 938
3 Kaufman M J, Eades J A, Loretto M H and Fraser H L 1983 *Met. Trans.* **14A** 1561
4 Donachie M J Jr and Kriege O H 1972 *J. Mater.* **7** 269
5 Sims C T and Hagel W C (ed) 1972 *The Superalloys: Vital High Temperature Gas Turbine Materials for Aerospace and Industrial Power* (New York: Wiley)

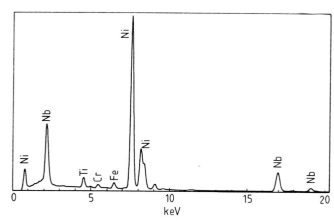

Figure 4.1 EDX spectrum from a particle of δ phase found in a specimen of IN 718. Note that it has apparently the same composition as the γ'' in that alloy.

DELTA

0001/010

11$\bar{2}$6/140

11$\bar{2}$3/120

2$\bar{1}$$\bar{1}$6/182

10$\bar{1}$2/382

10$\bar{1}$1/342

2$\bar{1}$$\bar{1}$3/142

1$\bar{1}$02/021

1$\bar{1}$01/011

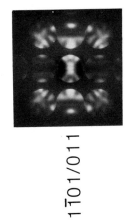

Figure 4.2 This figure is a partial ZAP map for δ phase. The map extends for approximately 45° from the [010] zone axis in the direction of the [001] and [100] axes. If the crystal were primitive hexagonal instead of primitive orthorhombic, then the axes labelled with equivalent hexagonal indices would be identical. Some of these patterns are quite similar, but it is clear that the crystal could not be taken as hexagonal. Patterns recorded at 120 kV.

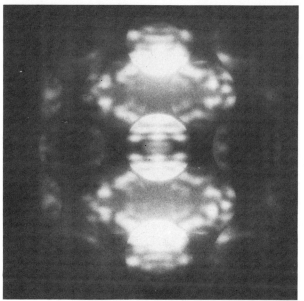

1 $\bar{1}$00/001

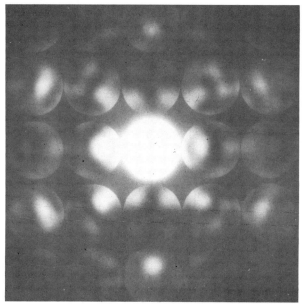

2 $\bar{1}$ $\bar{1}$0/102

010 TILT AXIS

11 $\bar{2}$0/100

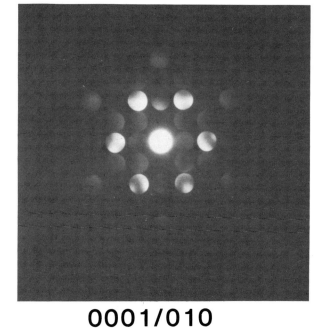

0001/010

001 TILT AXIS

Figure 4.3 Four major axes from the δ phase. The axes have been labelled with both hexagonal and orthorhombic indices. The reason for this can be inferred from the [0001]/[010] axis. The pattern has a hexagonal appearance when obtained from a rather thin region of the specimen. The deviation from hexagonality has been discussed in the previous figure. The alternate absent reflections in the vertical row through the origin in the [010] axis are due to the *n* glide which is parallel to the beam. Tilting by 90° about the [001] axis results in the [100] axis. The dark bars in the alternate vertical reflections ($\pm$010) are attributed to the *n* glide. If the crystal is then tilted about the [010] axis the [102] axis results. This axis is of particular interest. A low camera length view is illustrated below. The long camera length view above shows a system of concentric rings in the bright field disc with a bright centre. This is characteristic of a simple string axis close to a critical voltage (see Introduction). The odd $0k0$ reflections are kinematically rather than dynamically forbidden at this axis and there are strong double diffraction routes for these reflections, as can be seen in this figure, but more particularly in figure 4.4. Note that there is very little intensity variation along the $0k0$ row of reflections in figure 4.4. Tilting again about the [010] axis leads to the [001] axis. The odd $0k0$ reflections at this axis have no strong double diffraction routes because the $hk0$ reflections are kinematically forbidden when $h+k$ is odd. As a result the reflections are virtually absent. Patterns recorded at 120 kV.

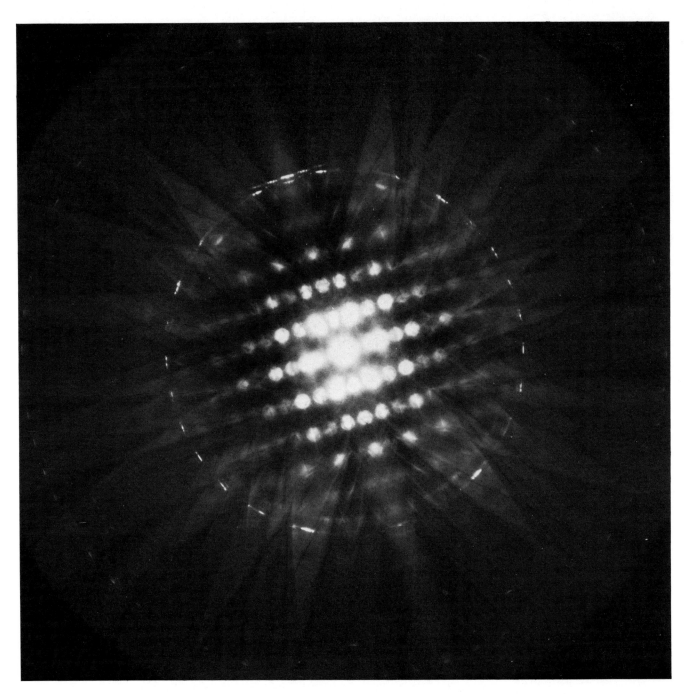

Figure 4.4 The $\langle 102 \rangle$ zone axis at low camera length. The HOLZ rings are interesting because only the first and the fourth are clearly visible. The second and third rings are essentially absent because of small structure factors associated with these zones. The pattern symmetry is *m*. Patterns recorded at 120 kV.

Phase name: σ phase
Lattice type: primitive tetragonal
Point group: 4/mmm
Space group: P4₂/mnm

Lattice parameters: $a = 0.8790$ nm, $c = 0.4544$ nm.
Values for FeCr [1]
$a = 0.9188$ nm, $c = 0.4812$ nm.
Values for FeMo [1]

σ

FORMULA AND COMPOSITIONAL VARIABILITY

There are a variety of σ phases in the transition metal systems. For an early review see Hall and Algie [1]. Typical σ compositions involved two principal transition metals (Fe and Cr or Fe and Mo) with a number of minor elements. The question of ordering in binary compounds was examined in two early references [2,3]. The extent of ordering in commercial alloys is uncertain, although atoms such as Mo segregate to the $8(j)$ sites and a thorough study has been made in two cases [4].

MORPHOLOGY

It is a hard, brittle phase which forms as large blocky precipitates at high-energy interfaces, especially at grain boundaries and triple points [5].

Certain orientation relationships have been observed frequently between the austenite matrix in stainless steels and the σ phase precipitates, e.g.

$(111)_\gamma//(001)_\sigma$; $\langle 011 \rangle_\gamma//\langle 140 \rangle_\sigma$ [5]
$(111)_\gamma//(001)_\sigma$; $\langle \bar{1}10 \rangle_\gamma//\langle \bar{1}10 \rangle_\sigma$ [6].

These relationships are very nearly equivalent.

OCCURRENCE

σ phase is a thermodynamically stable phase in 316 steel whose appearance is retarded by kinetic factors. It often appears after long aging or service times at temperatures of 600–700 °C but the likelihood of its formation depends on the amount of C, N, Mo and Si present (see below). It is likely to occur at earlier times if higher aging temperatures are used. Carbon in the matrix retards the formation of σ phase as does nickel. The presence of molybdenum and silicon promotes its precipitation. Cold working prior to aging or service has been shown to accelerate the formation of σ.

OTHER COMMENTS AND FEATURES

Raising the solution treatment temperature of the parent matrix tends to reduce the size of the σ particles when they form, but does not seem to affect the time to nucleation. The formation sequence of σ phase is not clearly established but Weiss and Stickler [7] postulated that it forms on dissolving $M_{23}X_6$ precipitates or on sites where $M_{23}X_6$ had previously existed. When σ particles, extracted onto a carbon replica, are heated with an intense electron beam an explosive reaction produces M_7X_3 [8] (see § 11).

REFERENCES

1 Hall E O and Algie S H 1966 *Met. Rev.* **11** 61
2 Kasper J S and Waterstrat R M 1956 *Acta Crystallogr.* **9** 289
3 Wilson C G and Spooner F J 1963 *Acta Crystallogr.* **16** 230
4 Yakel H L 1983 *Acta Crystallogr.* B **39** 20
5 Lewis M H 1966 *Acta Met.* **14** 1421
6 Beckitt F R 1969 *J. Iron Steel Inst.* **207** 632
7 Weiss B and Stickler R 1972 *Met. Trans.* **3A** 851
8 Cooke K E 1977 *Developments in Electron Microscopy and Analysis 1977* (Inst. Phys. Conf. Ser. 36) p 431
9 Self P G, O'Keefe M A and Stobbs W M 1983 *Acta Crystallogr.* B **39** 197

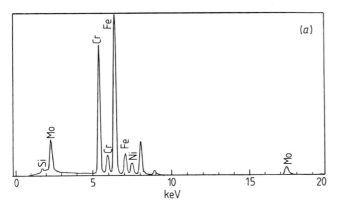

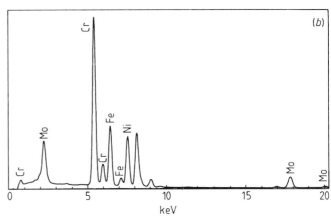

Figure 5.1 Characteristic EDX spectra from σ phases found in (*a*) 316 steel and (*b*) IN 718.

Figure 5.2 (*right*) CBED ZAP map for σ phase in 316 steel.

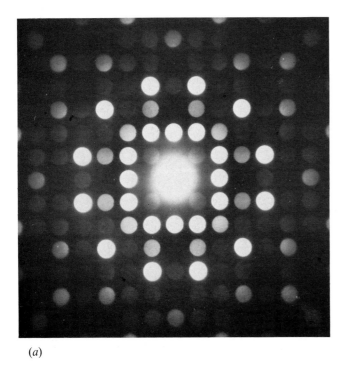

(*a*)

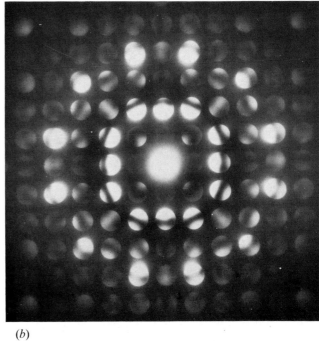

(*b*)

Figure 5.3 σ phase [001] ZAPs from (*a*) thin and (*b*) thick crystals using a small condenser aperture. In (*b*) the dark bars in the odd *h*00 and 0*k*0 reflections can be seen. These absences are due both to glide planes parallel to the zone axis and screw axes perpendicular to it. The reflections are not visible for the thinner crystal. Note that the ring of 'dark bars' in the second order dark field discs are not dynamic absences and should not be confused with them. It has aroused some recent interest, see [9].

28

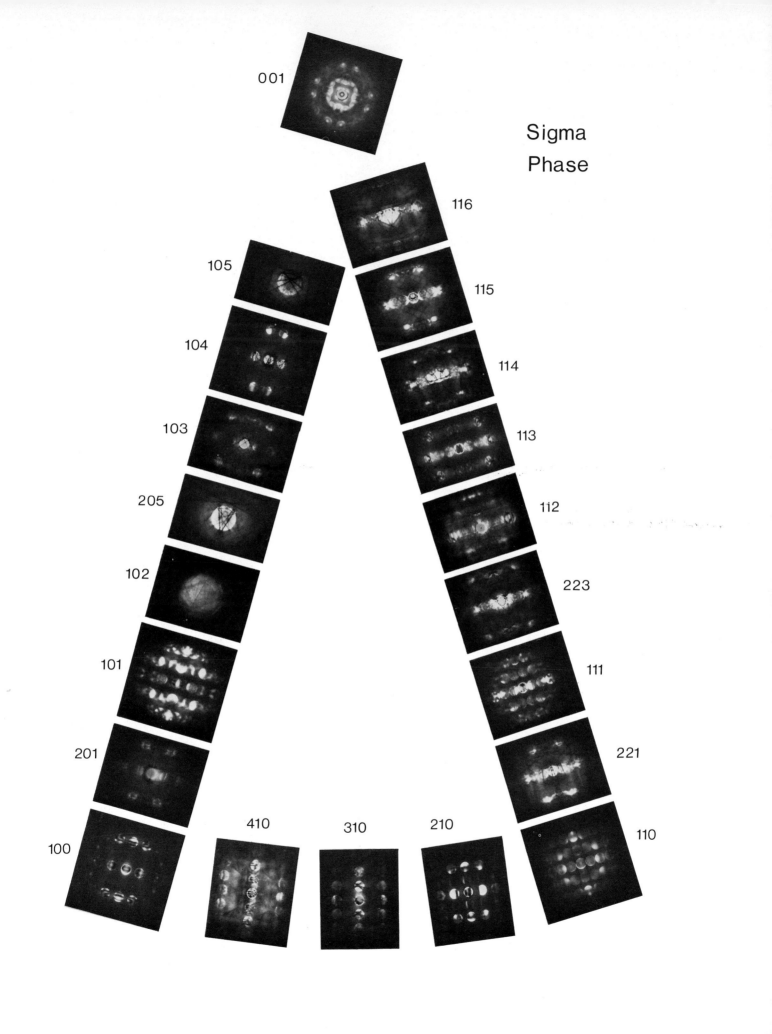

Sigma Phase

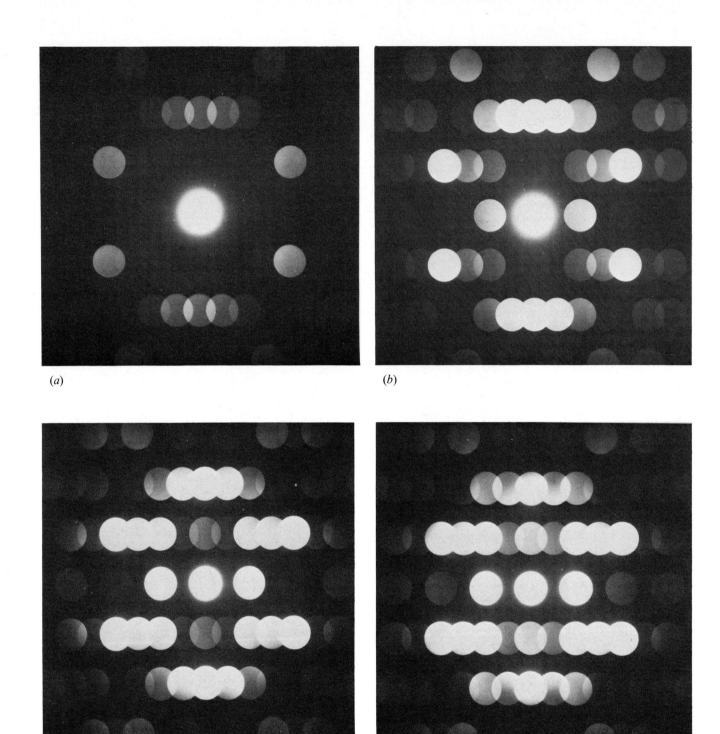

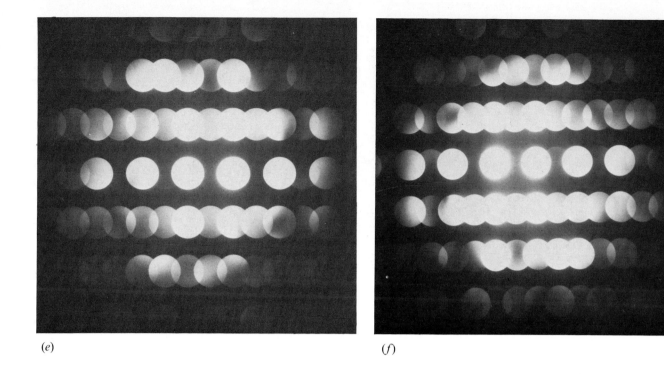

(e)

(f)

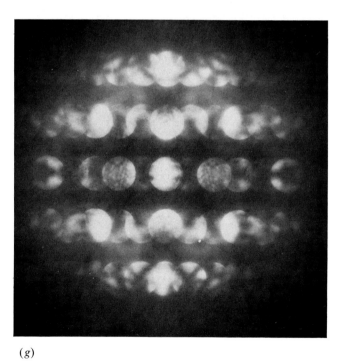

(g)

Figure 5.4 σ phase ⟨011⟩ thickness sequence. Patterns from the thinnest crystal show only the *2mm* projection symmetry. The thickest region shows HOLZ information and reveals that the whole pattern symmetry is in fact only *m*. This identifies the diffraction group as $2_R mm_R$.

Figure 5.5 σ ⟨102⟩ axis. This is illustrated because, although not a major axis, it is a characteristic one. The zero-layer diffracted discs are very weak but many of the FOLZ reflections are strong. As a result, a thick piece of crystal leads to absorption of most of the zero-layer dark field discs leaving only the bright field disc.

As the bright field pattern can be observed with minimal interference from the dark fields a very large condenser aperture can be used (all the high symmetry σ ZAPS require a small aperture to avoid disc overlap). The pattern contains virtually no thickness fringes and approximately one hundred HOLZ lines.

Phase name: Laves phase (sometimes termed 'η phase')
Lattice type: hexagonal, isostructural with $MgZn_2$
Point group: $6/mmm$
Space group: $P6_3/mmc$
Lattice parameters: $a = 0.4730$ nm, $c = 0.7720$ nm in 316 stainless steel [1]
$a = 0.4744$ nm, $c = 0.7725$ nm in Mo-containing alloys [2]

Laves or η

FORMULA AND COMPOSITIONAL VARIABILITY

The chemical constitution of this phase takes the form M_2M' where in 316 steel M = Cr, Fe, Ni and M' = Mo or W [1]. In superalloys M = Co, Cr, Fe, Ni and M' = Nb, Mo or Ta [3].

There is evidence (but not general agreement [4,5]) that Laves phase has some solubility for carbon and that the introduction of carbon into the lattice produces the heavy faulting that is frequently observed in this phase. Laves phase also forms in a variety of polytypes [7,8].

MORPHOLOGY

Laves phase typically occurred as large (> 1 μm) particles on grain boundaries. In 316 steel it was often found in close association with grain boundary $M_{23}X_6$. The fault density was very high and sometimes occurred on more than one set of crystallographic planes.

Different orientation relationships between the matrix and Laves have been noted,

e.g. for Fe_2Nb in a Nb-containing alloy
$(0001)_L//(111)_\gamma$; $(1\bar{1}00)_L//(\bar{1}10)_\gamma$ [6]
and for Fe_2Mo in an austenitic steel
$(10\bar{1}3)_L//(111)_\gamma$; $(\bar{1}2\bar{1}0)_L//(\bar{1}2\bar{1})_\gamma$ [6]
where L = Laves and γ = matrix.

OCCURRENCE

Laves phase forms in alloys after prolonged thermal aging. A typical example would be that of 316 steel after approx 10 000 hr at 600 °C. There is some dependence of the formation of the phase upon prior heat treatment.

OTHER FEATURES AND COMMENTS

Precipitates of Laves phase on carbon extraction replicas are sensitive to beam heating and can react vigorously to form M_7X_3.

REFERENCES

1 Weiss B and Stickler R 1972 *Met. Trans.* **3A** 851
2 Bechtoldt C J and Vacher H C 1957 *J. Res. Nat. Bur. Stand.* **58** 7
3 Sims C T and Hagel W C (ed) 1972 *The Superalloys: Vital High Temperature Gas Turbine Materials for Aerospace and Industrial Power* (New York: Wiley)
4 Wiegand H and Doruk M 1962 *Arch. Eisen.* **33** 559
5 Thier H, Baumel H A and Schmidtman E 1969 *Arch. Eisen.* **40** 333
6 Denham A W and Silcock J M 1969 *J. Iron Steel Inst.* **207** 585
7 Allen C W, Delavignette P and Amelinckx S 1972 *Phys. Stat. Sol.* a **9** 237
8 Komura Y and Kitano Y 1977 *Acta Crystallogr.* B **33** 2496

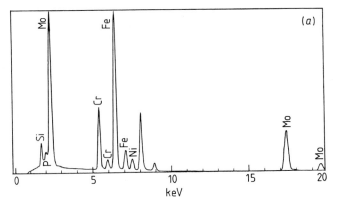

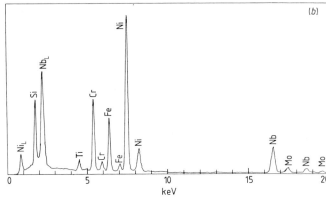

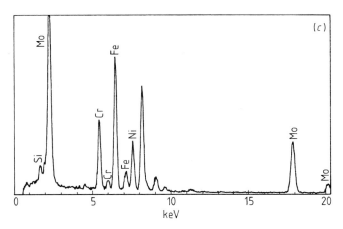

Figure 6.1 EDX spectra from Laves phase particles discovered in (*a*) 316 steel, (*b*) Inconel 718 and (*c*) Incoloy 901.

Figure 6.2 (right) CBED ZAP map of Laves phase studied in 316 steel.

(*a*)

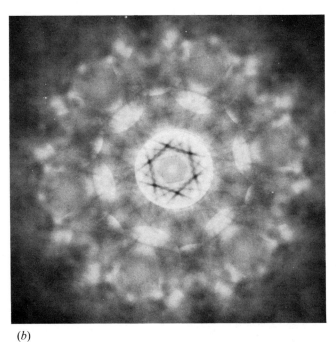

(*b*)

Figure 6.3 (*a*) Intermediate camera length and convergence angle [0001] pattern. (*b*) Long camera length and larger convergence angle [0001] pattern. In both cases the crystal was rather thick and diffuse scattering has reduced the disc visibility.

34

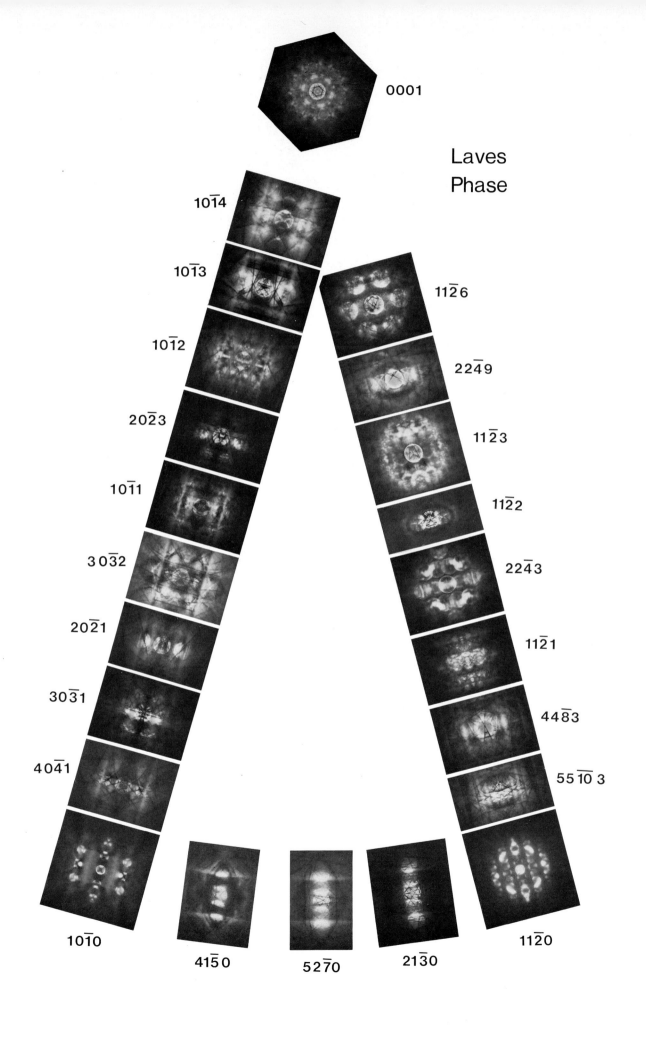

0001

Laves
Phase

10$\bar{1}$4

10$\bar{1}$3

10$\bar{1}$2

20$\bar{2}$3

10$\bar{1}$1

3 0$\bar{3}$2

20$\bar{2}$1

30$\bar{3}$1

40$\bar{4}$1

10$\bar{1}$0

41$\bar{5}$0

52$\bar{7}$0

21$\bar{3}$0

11$\bar{2}$6

22$\bar{4}$9

11$\bar{2}$3

11$\bar{2}$2

22$\bar{4}$3

11$\bar{2}$1

44$\bar{8}$3

55$\overline{10}$ 3

11$\bar{2}$0

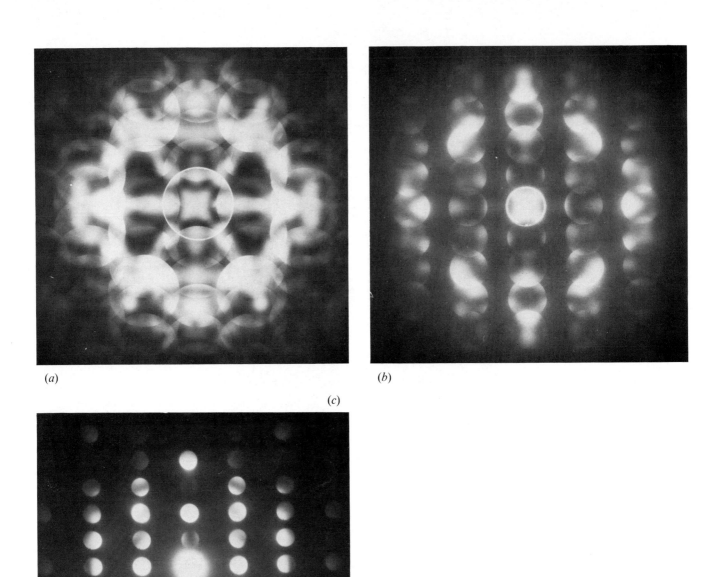

(a)

(b)

(c)

Figure 6.4 ⟨11$\bar{2}$0⟩ Laves ZAPs selected to illustrate the change in appearance with change of convergence angle (second condenser aperture). (*a*) Large aperture. (*b*) Intermediate aperture. (*c*) Small aperture, revealing dynamic absences in the vertical row through the centre.

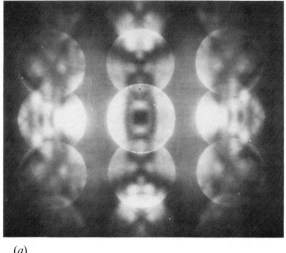

(a)

(b)

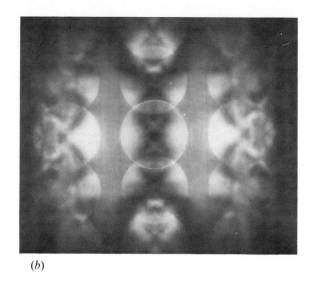

(c)

Figure 6.5 The $\langle 1\bar{1}00 \rangle$ axis of Laves phase is one that changes significantly but not dramatically with change of thickness. The patterns (a) to (c) are in order of increasing thickness.

Figure 6.6 Two zone axes of Laves phase which have been selected for their interesting features. (a) Inelastic scattering almost obliterates the disc and dark field detail of this $\langle 2\bar{1}\bar{1}3 \rangle$ pattern; nevertheless, the pattern mirror symmetry is still readily determined. Note that the detail of such a pattern is often scarcely visible in the microscope. It is therefore important to record ZAPs photographically to study the symmetry. (b) The $\langle 30\bar{3}2 \rangle$ pattern has an approximately square arrangement of reflections and could lead to confusion with cubic or tetragonal phases. Note, however, that the pattern only has a single mirror line.

(a)

(b)

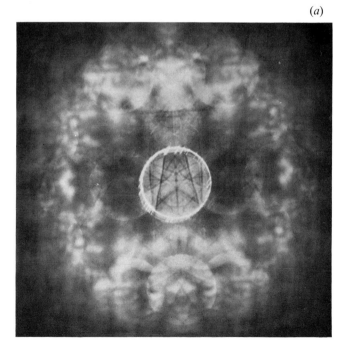

χ

Phase name: χ phase
Lattice type: bcc (α Mn type) [1]
Point group: $\bar{4}3m$
Space group: $I\bar{4}3m$
Lattice parameters: $a = 0.8862$–0.8878 nm in 316 steel [2,3]
$a = 0.8920$ nm in the Fe, Cr, Ni, Mo system of Kasper [1]

FORMULA AND COMPOSITIONAL VARIABILITY

χ is an intermetallic phase with 58 atoms in its unit cell. In stainless steels it is generally composed of Fe, Cr and Mo. Goldschmidt [4] states that the phase is capable of incorporating carbon and can be considered as a carbide (formula $M_{18}C$) or an intermetallic. The stoichiometry of the intermetallic has been determined to be:

$Fe_{36}Cr_{12}Mo_{10}$ [1]
$(Fe,Ni)_{36}Cr_{18}Mo_4$ [2].

MORPHOLOGY

The morphology of this phase varies from rod shapes to large globular particles similar to σ. The particles form successively on grain boundaries, incoherent twin boundaries and then intragranularly. After cold working and aging, intragranular rods form which are coherent with the matrix. Particles often form in the vicinity of carbide precipitates. The orientation relationship between the matrix and the χ particles may be written: $(111)_\gamma//(110)_\chi$; $\langle01\bar{1}\rangle_\gamma//\langle\bar{1}10\rangle_\chi$.

OCCURRENCE

χ phase forms during high-temperature aging of steels. In 316 steel the precipitation time is not affected by changes in solution treatment temperature. Cold working prior to aging, however, increases the amount of intragranular χ phase that is formed.

OTHER COMMENTS AND FEATURES

The ratio of the quantity of χ phase formed to the quantity of σ phase formed is decreased by cold working prior to aging. Unlike σ or Laves phases no beam heating reaction of χ phase and CERs has been observed.

REFERENCES

1 Kasper J S 1954 *Acta Met.* **2** 456
2 Kautz H R and Gerlach H 1968 *Arch. Eisen.* **2** 151
3 Andrews K W 1949 *Nature* **164** 1015
4 Goldschmidt H J 1967 *Interstitial Alloys* (London: Butterworth)

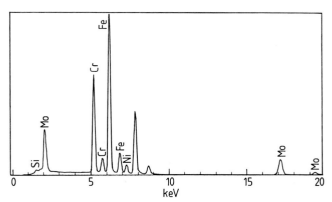

Figure 7.1 Characteristic EDX spectrum from χ phase found in 316 steel.

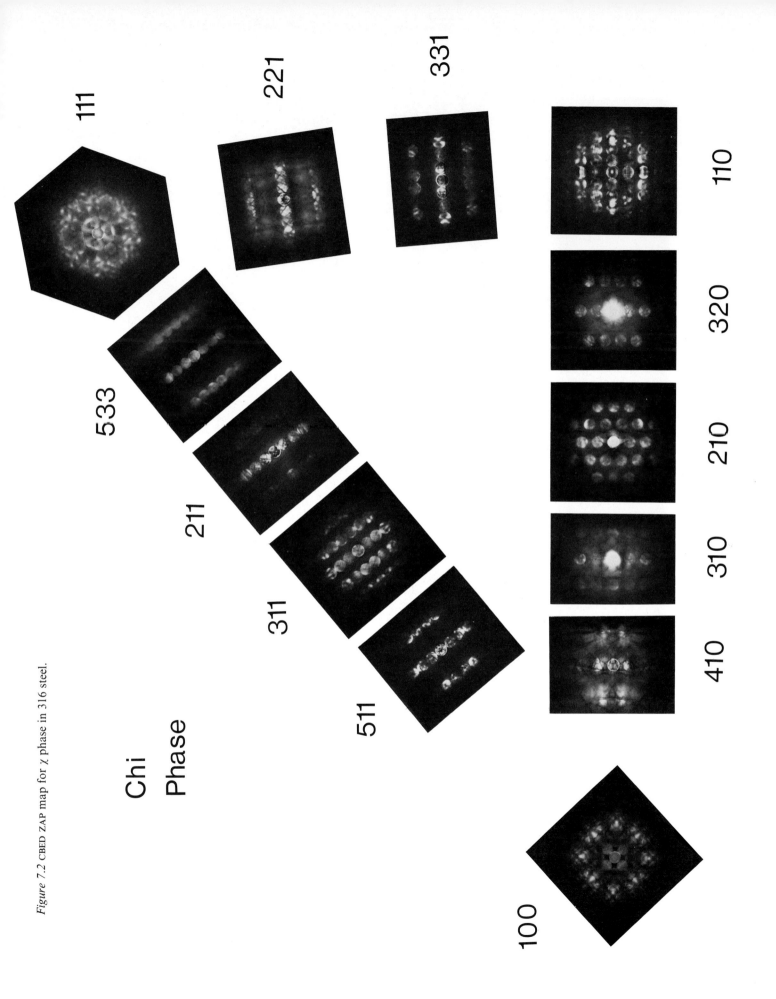

Figure 7.2 CBED ZAP map for χ phase in 316 steel.

Chi Phase

111

221

331

110

533

211

311

511

320

210

310

410

100

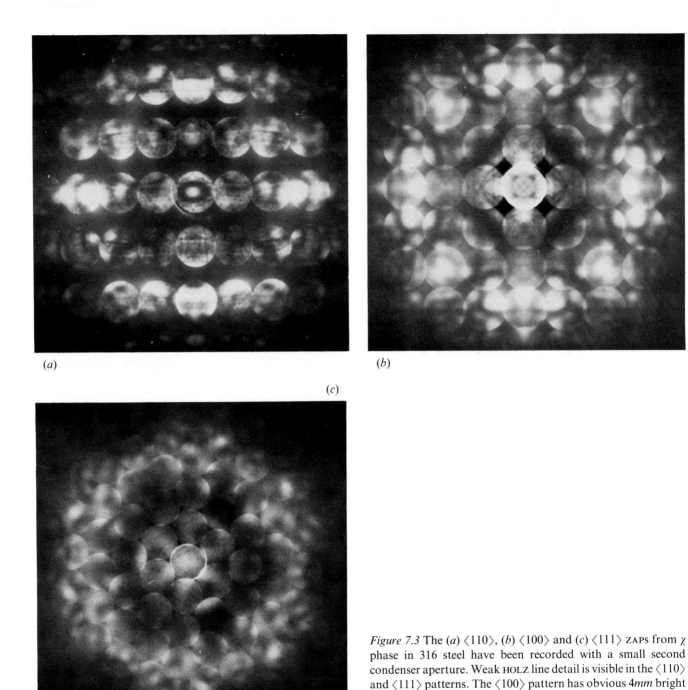

(a)

(b)

(c)

Figure 7.3 The (a) ⟨110⟩, (b) ⟨100⟩ and (c) ⟨111⟩ ZAPS from χ phase in 316 steel have been recorded with a small second condenser aperture. Weak HOLZ line detail is visible in the ⟨110⟩ and ⟨111⟩ patterns. The ⟨100⟩ pattern has obvious 4*mm* bright field symmetry but the HOLZ contrast reduces the symmetry to 2*mm*. The ⟨111⟩ pattern shows 3*m* whole pattern symmetry.

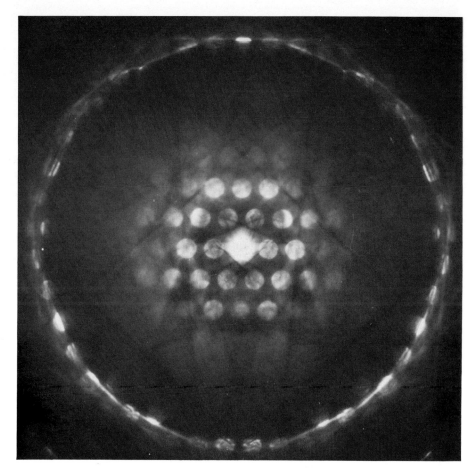

Figure 7.4 ⟨210⟩ χ phase CBP. Although this pattern at first appears to contain a mirror line, close examination of the FOLZ ring reveals that the pattern symmetry is in fact 1.

$M_{23}X_6$

Phase name: $M_{23}X_6$
Lattice type: fcc
Point group: $m3m$
Space group: $Fm3m$
Lattice parameters: $a = 1.056–1.065$ nm [1]

FORMULA AND COMPOSITIONAL VARIABILITY

The structure contains 116 atoms in the unit cell (92 metal atoms and 24 carbon atoms) [2]. It is essentially $Cr_{23}C_6$ with partial substitution of Cr by other metals. Weiss and Stickler [1] reported an empirical formula $Cr_{16}Fe_5Mo_2C_6$ for the precipitate in 316 stainless steel. Many workers have noted that the carbon may be substituted by nitrogen, boron and phosphorus [3, 4, 5]. Hence, a general formula is employed here. Note the similarity with G-phase [6,7] which can be distinguished by its Si content and by its different lattice parameter range.

MORPHOLOGY

This is very varied and a wide range is on record, including irregular shapes, needles, plates, thin sheets and octahedra.

OTHER COMMENTS AND FEATURES

The most frequently observed orientation relationship between the $M_{23}X_6$ and the parent matrix has been determined as:
$$\{100\}_{M_{23}X_6} / / \{100\}_\gamma; \ \langle 100 \rangle_{M_{23}X_6} / / \langle 100 \rangle_\gamma \ [8,9].$$

OCCURRENCE

It is usually the first and often the only carbide to form in a wide range of steels at low service temperatures and in the absence of strong carbide formers such as Nb, W, Ti and V. The phase forms, in order of increasing difficulty, at high-angle grain boundaries, incoherent twin boundaries and coherent twin boundaries. It also forms intragranularly.

REFERENCES

1 Weiss B and Stickler R 1972 *Met. Trans.* **3A** 851
2 Bowman A L, Arnold G P, Storms E K and Nereson N G 1972 *Acta Crystallogr.* B **28** 3102
3 Matsuo T, Shinoda T and Tanaka R 1973 *Tetsu to Haganē* **69** 907
4 Williams T M and Talks M G 1972 *J. Iron Steel Inst.* **210** 870
5 Maitrepierre Ph, Thivillier D and Tricot R 1975 *Met. Trans.* **6A** 287
6 Beattie H J Jr and VerSynder F L 1956 *Nature* **178** 208
7 Goldschmidt H J 1967 *Interstitial Alloys* (London: Butterworth)
8 Lewis M H and Hattersley B 1965 *Acta Met.* **13** 1159
9 Beckitt F R and Clarke B R 1967 *Acta Met.* **15** 113

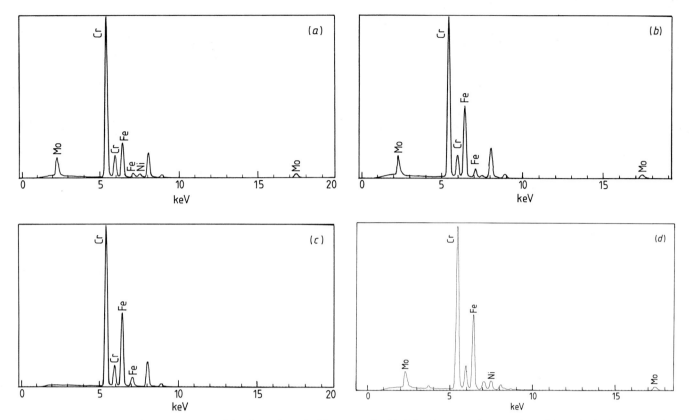

Figure 8.1 EDX spectra from $M_{23}X_6$ particles in 316 steel. The first three spectra represent the three characteristic types discovered in this material. (*a*) is the most common type discovered and shows Cr, Fe and Mo with a trace of Ni. (*b*) is similar to (*a*) except for a larger Fe content. (*c*) does not contain any Mo. The last spectrum (*d*) was recorded from a particle embedded in the austenite matrix. Although the iron peak is larger than in the other spectra (matrix contribution), the spectrum can still be identified as that of $M_{23}X_6$ by the relative intensities of the other peaks. The $M_{23}X_6$ that occurred in Astroloy had an almost identical spectrum to (*a*) above.

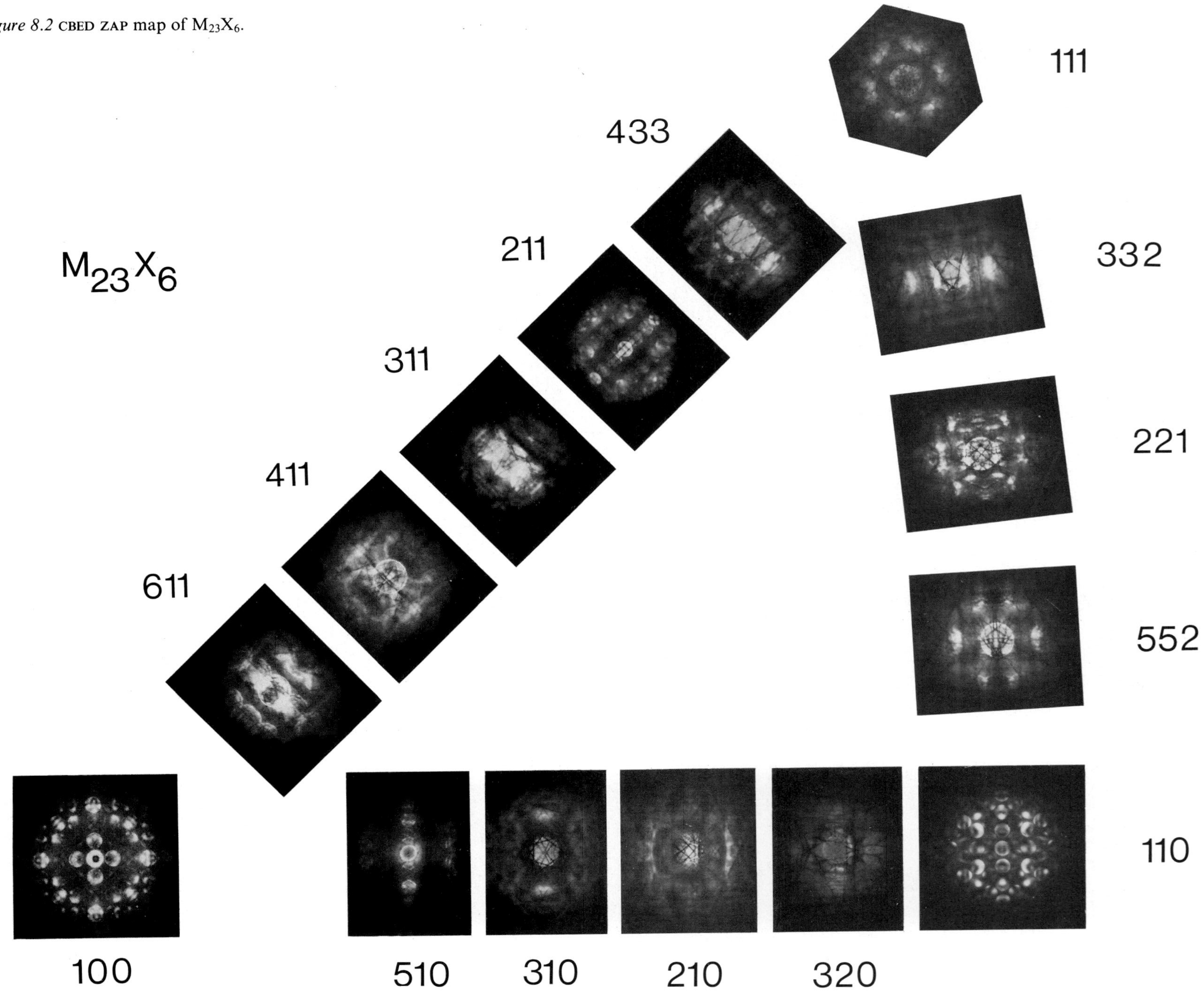

Figure 8.2 CBED ZAP map of $M_{23}X_6$.

Figure 8.3 ⟨110⟩ $M_{23}X_6$ pattern. The specimen was so thin that no detail is visible in the discs. However, the relative diffracted intensities make the pattern distinctive.

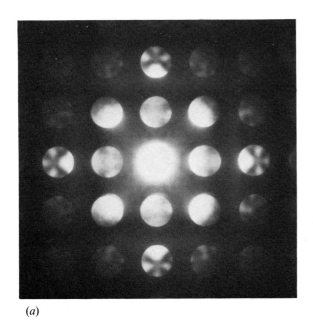

(a)

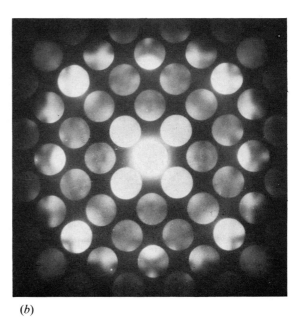

(b)

Figure 8.4(a) ⟨100⟩ CBP from $M_{23}X_6$. (b) ⟨100⟩ CBP from M_6X reproduced to the same scale as (a) to facilitate comparison. Unless care is taken these two patterns can be confused because the kinematically forbidden 002 spots in M_6X become quite strong by double diffraction routes via the FOLZ if the crystal is not accurately aligned on the ⟨100⟩ axis. It is easy to confuse the patterns using conventional selected area diffraction.

(a)

(b)

(c)

(d)

Figure 8.5 Four of the major axes of $M_{23}X_6$ recorded at longer camera lengths than on the ZAP map. (a) ⟨100⟩. A large second condenser aperture has been used here to reveal a considerable amount of HOLZ detail in the bright field disc. (b) ⟨111⟩. Use of a large second condenser aperture produced overlapping discs but yielded evidence of 3*m* symmetry in the central beam. This pattern should be compared with the ⟨111⟩ pattern for M_6X (figure 9.3). The patterns, although complicated, are clearly distinct. (c) ⟨110⟩. The use of a small second condenser aperture allowed the individual discs to be resolved. The HOLZ contrast is almost invisible in this pattern. (d) ⟨211⟩. A small aperture has been used so that the individual discs could be seen. HOLZ detail is visible in the pattern.

46

Phase name: M_6X also known as 'η-carbide'
Lattice type: fcc
Point group: $m3m$
Space group: $Fd3m$
Lattice parameters: $a = 1.095$ nm in 316 steel [1]. In general a range of values from 1.07 to 1.22 nm is quoted depending on composition.

FORMULA AND COMPOSITIONAL VARIABILITY

The phase is frequently described as M_6C; however, evidence [2] indicates that it may sometimes be $M_3M'_3X$, $M_6M'_6X$ (where X = C, N or O) or $M_2M'_3SiX$ (with X sites vacant and Si on metal sites) [2]. The phase does not have a strict stoichiometry. Although its composition depends on the matrix composition, it is very unusual amongst carbide precipitates in steel in having a high nickel content (cf γ' and γ''). Nickel is not noted for its carbide formation [3]. Phosphorus may also occupy the X sites.

MORPHOLOGY

The morphology is variable and almost indistinguishable from $M_{23}X_6$ apart, perhaps, from a greater tendency for faceting.

OCCURRENCE

This phase is common in steels after prolonged aging. It nucleates on $M_{23}X_6$ in a cube–cube orientation [4,5]. In 316 steel it tends to form near the surface when nitrogen diffusion from the atmosphere is evident. Thier *et al* [4] noted that N_2 accelerates the formation of M_6X. Silicon, entering the lattice on M atom sites in tool steels, is said to reduce the cell size and stabilise the phase.

REFERENCES

1 Weiss B and Stickler R 1972 *Met. Trans.* **3A** 851
2 Stadelmaier H H 1969 Metal Rich Metal–Metalloid Phases in *Developments in the Structural Chemistry of Alloy Phases* ed B C Giessen (New York: Plenum Press) p 141
3 Kubashewski O and Evans E L L 1956 *Metallurgical Thermochemistry* (London: Pergamon)
4 Thier H, Baumel H A and Schmidtman E 1969 *Arch. Eisen.* **40** 333
5 Mukerjee T and Dyson D J 1972 *J. Iron Steel Inst.* **210** 203

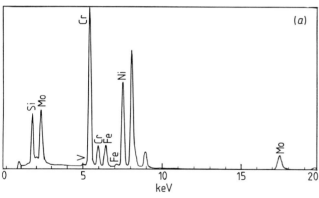

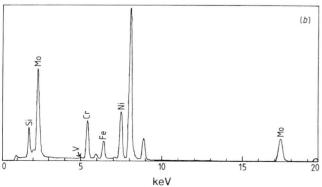

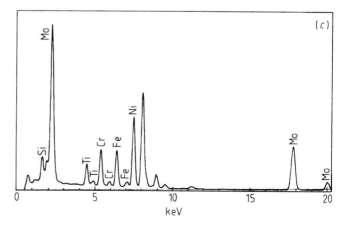

Figure 9.1 EDX spectra from M_6X precipitates in two samples of 316 steel (*a*) and (*b*) and a sample of IN 901 (*c*). This phase exhibits a large degree of variation in its composition and can probably be identified reliably only by diffraction. However, the spectra above represent commonly occurring compositions. Note the trace of V in (*a*) and (*b*) and the silicon content.

Figure 9.2 CBED ZAP map for M$_6$X discovered in 316 steel.

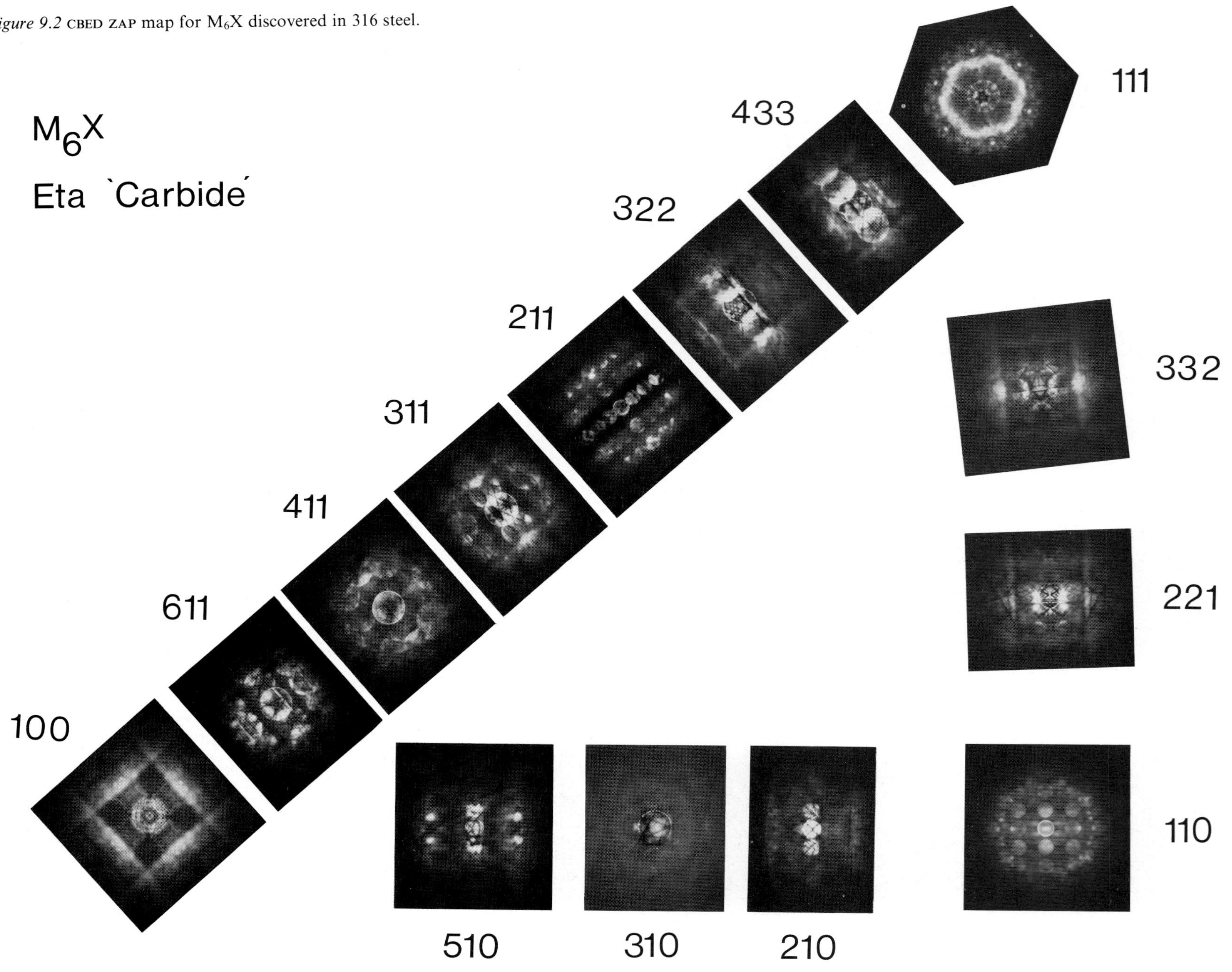

M$_6$X

Eta 'Carbide'

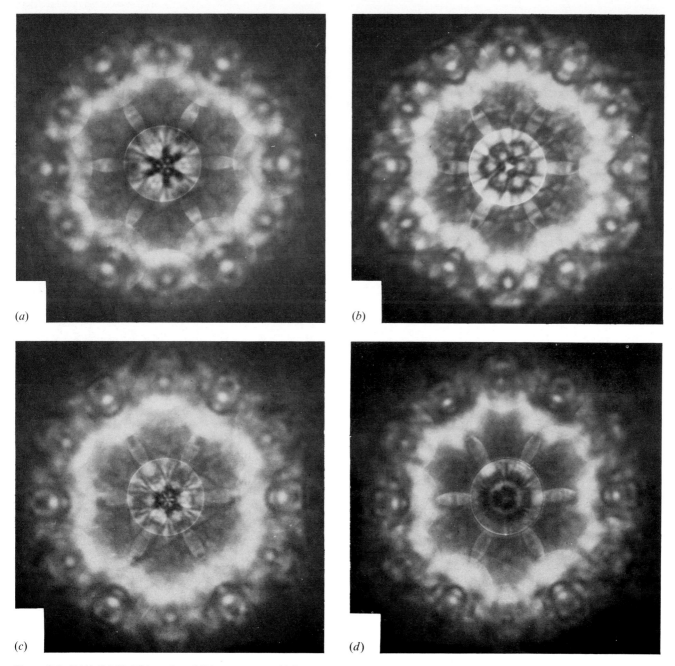

(a)

(b)

(c)

(d)

Figure 9.3 ⟨111⟩ M₆X. This axis exhibits patterns which vary considerably with thickness. The four patterns shown are in order of increasing thickness. Use of a large condenser aperture has resulted in overlapping discs.

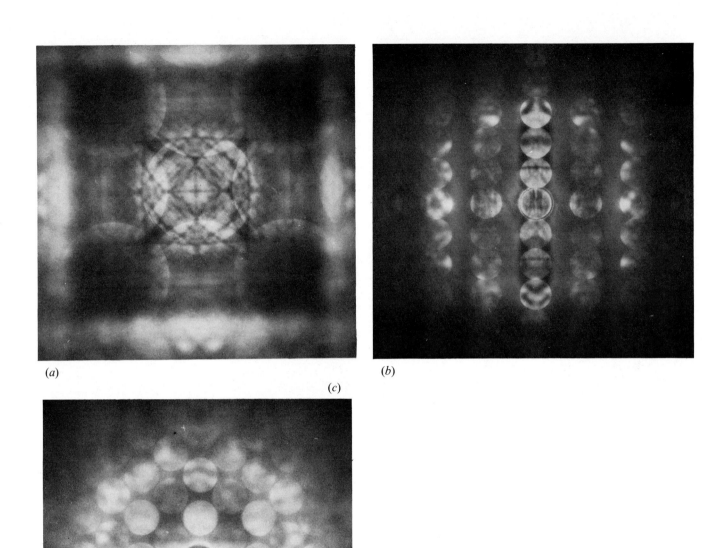

(a)

(b)

(c)

Figure 9.4 Three other major zone axes from M_6X. (a) $\langle 100 \rangle$. A large condenser aperture has been used here. The discs overlap, but the HOLZ detail is almost continuous because the crystal was rather thick and the effects of diffuse scattering are making a large contribution to the image. The bright field disc displays the expected 4*mm* symmetry. (b) $\langle 211 \rangle$. A diffraction pattern from quite a thick crystal using a small condenser aperture which allows HOLZ detail to be seen in the individual discs. The pattern symmetry contains only a single mirror. (c) $\langle 110 \rangle$. The pattern contains no visible HOLZ detail in the central disc.

Phase name: MX, Hägg compounds [1]
Lattice type: fcc
Point group: $m3m$
Space group: $Fm3m$
Lattice parameters: The lattice parameters vary considerably with composition, as illustrated by the examples below. The Hägg compounds generally have a non-metal/metal atom radius ratio of less than 0.59.

Compound	a (nm)
$NbC_{0.99}$	0.4470 [2]
$NbC_{0.71}$	0.4431 [2]
TiC	0.4328 [3]
MoC	0.4274 [4]
TiN	0.4244 [3]
$VC_{0.87}$	0.4166 [3]
$VC_{0.73}$	0.4131 [3]

MX
Hägg compounds

OCCURRENCE AND MORPHOLOGY

In superalloys [5] MX can form during solidification (usually as large, thick, discrete particles) or it can form in subsequent heat treatments. These two different forms generally have different compositions. Primary (solidification) precipitates are usually intragranular. Heat treatment precipitates can be inter- or intragranular. In the work summarised in this text the size and morphology varied from alloy to alloy, but, most commonly, heat treatment precipitates occurred as disc-shaped particles. In Incoloy 901 they also appeared as intergranular laths and dendritic sheets.

In steels, MX is found when there are significant quantities of Nb, V or Ti added to the matrix, i.e. in stabilised steels. The morphology is quite varied: for example, NbC has been reported as approximately spherical particles, (V,Cr)C was found as thin plates and TiN was discovered as cruciform precipitates.

FORMULA AND COMPOSITIONAL VARIABILITY

M = Ti, V, Mo, Nb, Cr, Fe, W and Hf depending upon the composition of the parent alloy. The phase is generally of mixed composition when found in steels and superalloys. X = C and/or N and is quite frequently below stoichiometric proportions.

REFERENCES (for MX and MnS)

1 Toth L E 1971 *Transition Metal Carbides and Nitrides* (London: Academic)
2 Storms E K and Krickorian N H 1960 *J. Phys. Chem.* **64** 1471
3 Goldschmidt H J 1967 *Interstitial Alloys* (London: Butterworth)
4 Clougherty E V, Loltrop K H and Kaflas J A 1961 *Nature* **191** 1194
5 Sims C T and Hagel W C (ed) 1972 *The Superalloys: Vital High Temperature Gas Turbine Materials for Aerospace and Industrial Power* (New York: Wiley)
6 Kiessling R and Lange N 1978 *Non-metallic Inclusions in Steel* (London: The Metals Society)
7 Jones P, Rackham G M and Steeds J W 1977 *Proc. R. Soc.* A **354** 197

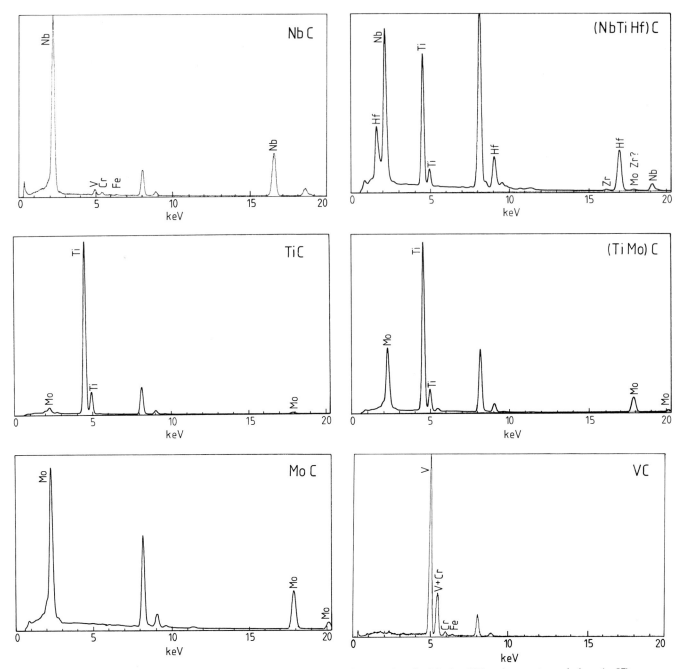

Figure 10.1 (a) EDX spectra from various MX particles found in the steels and superalloys studied for this text. The spectra can be correlated with the diffraction patterns below (p 57).

The structure of MX type carbides and nitrides is exactly the same as manganese sulphide, hence diffraction patterns from both MX precipitates and MnS inclusions are included in this section. However, since the sulphide has a different lattice parameter, frequency of occurrence and morphology from the carbides and nitrides, the details specific to MnS are included in the following subsection.

Phase name: manganese sulphide
Lattice parameter: $a = 0.5224$ nm [6]

FORMULA AND COMPOSITIONAL VARIABILITY

α manganese sulphide can incorporate considerable amounts of other transition metals in its lattice as well as Ca and Mg [6]. However, in our observations of 316 steel it was composed almost entirely of Mn and S. The Cr peak in the accompanying EDX spectrum is probably due to small particles of $M_{23}X_6$ that were frequently found in the vicinity of MnS.

OCCURRENCE AND MORPHOLOGY

It can occur as blocky particles in forged and moulded alloys. In wrought alloys it is usually found as long stringers (often 100 μm long) which lie in the rolling direction. The formation of MnS in Fe-based alloys is preferable to the grain boundary segregation of S where it can melt and cause mechanical failure during forming (hot or red shortness).

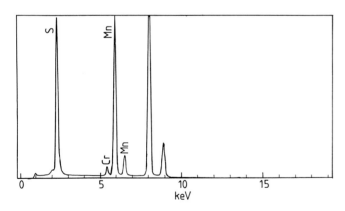

Figure 10.1 (*b*) A typical EDX spectrum from a MnS inclusion in a piece of 316 steel.

Figure 10.2(a) CBED ZAP map of NbC precipitation in a Nb stabilised steel.

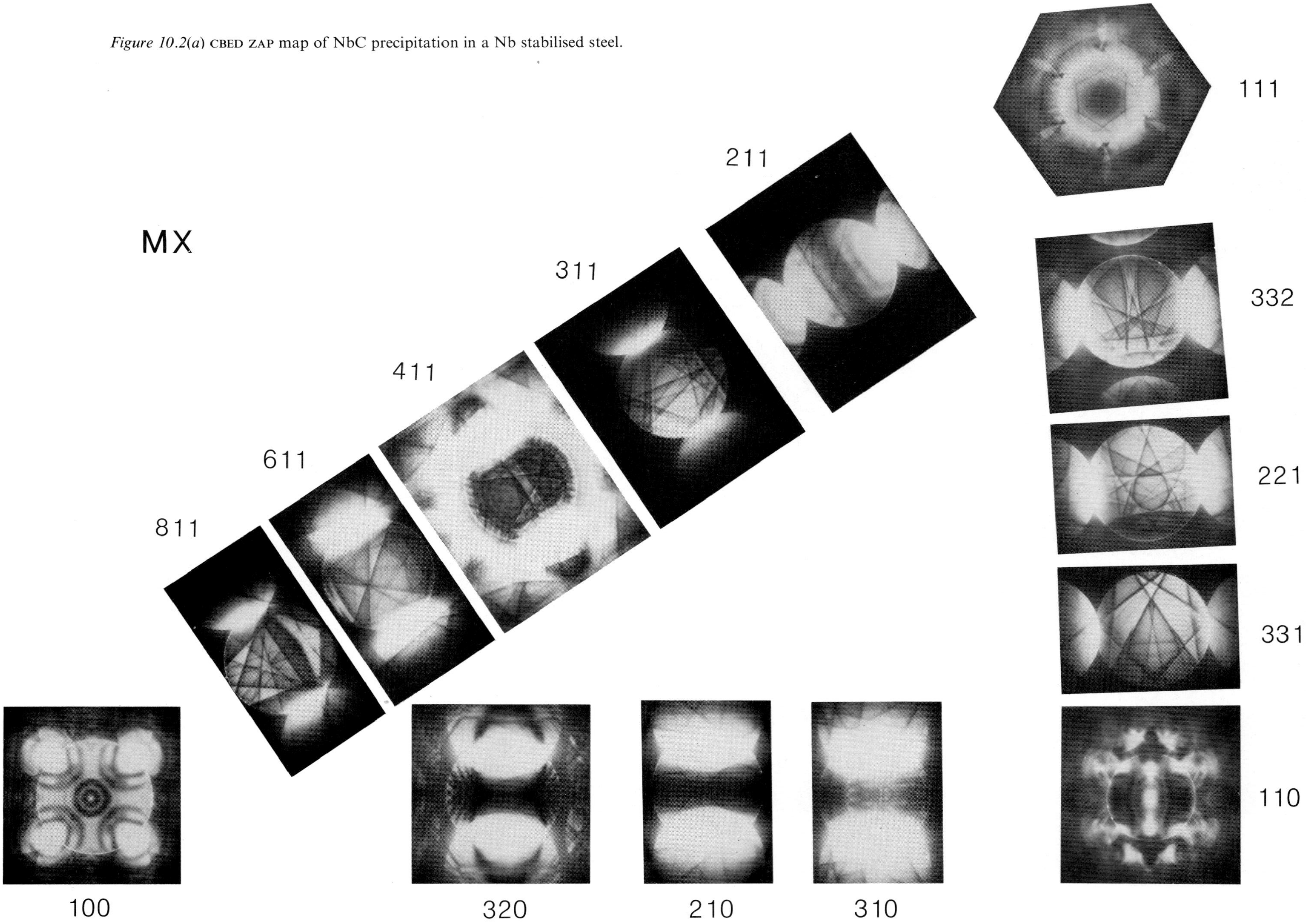

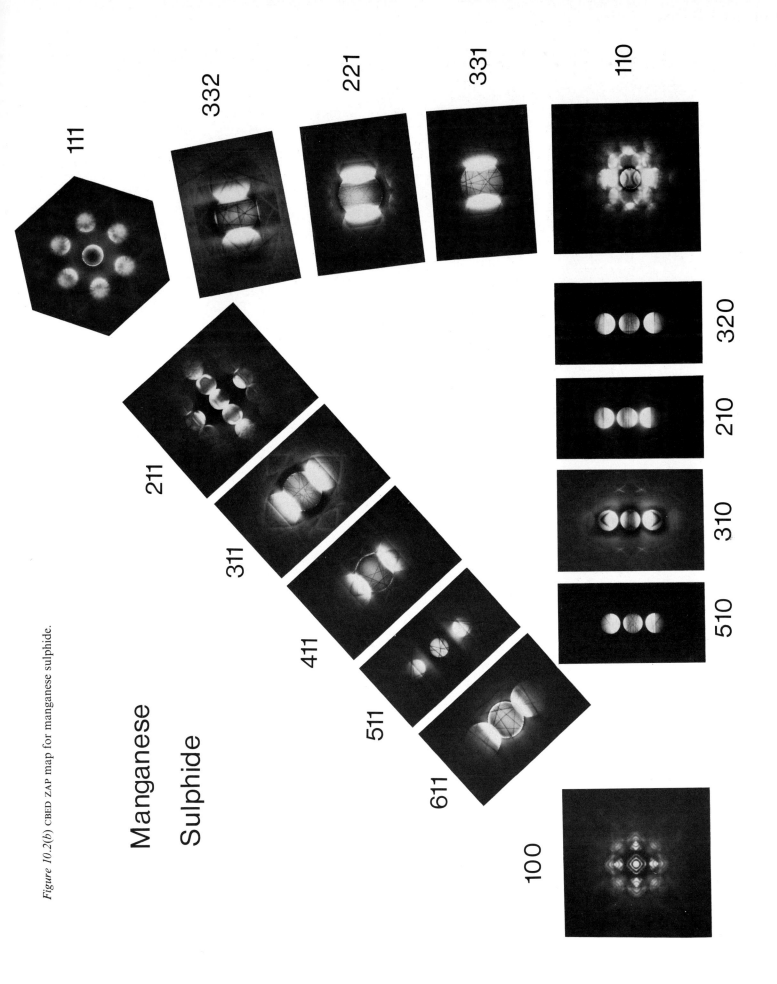

Figure 10.2(b) CBED ZAP map for manganese sulphide.

(a)

(b)

(c)

(d)

Figure 10.3 The ⟨100⟩ thickness sequence for MnS shows that 120 kV is near to, but below, a critical voltage for this simple string axis (see Introduction). The centre of the pattern always has a bright spot.

Figure 10.4 (*opposite page*) Six ⟨411⟩ axes from different MX precipitates in a number of steels and alloys. The particles are of different thicknesses and therefore the amount of fringe detail in the bright field varies. However it is not this but the HOLZ line detail which is of interest in these patterns. The off-stoichiometry characteristic of these phases generates lattice displacements which tend to degrade the large-angle scattering of high-energy electrons. As a result, HOLZ lines are not often visible at any of the low index axes (the NbC pattern in figure 10.5 is an exception) and it is necessary to go to the ⟨411⟩ axis to obtain reasonable patterns (for room temperature experiments).

As the lattice parameter (a_0) decreases (left to right and down the page) the HOLZ lines can be seen to shift. Note particularly the two almost horizontal lines across the centre of the bright field disc in NbC. With decreasing a_0 these two lines move towards one another, cross over (they are almost superimposed in the case of (Ti,Mo)C) and then move apart again. Note the approximate correlation between the changes evident in the patterns and the lattice parameters given in the table on page 51. For a quantitative comparison the precise compositions and stoichiometries would be required [7].

411 Axes

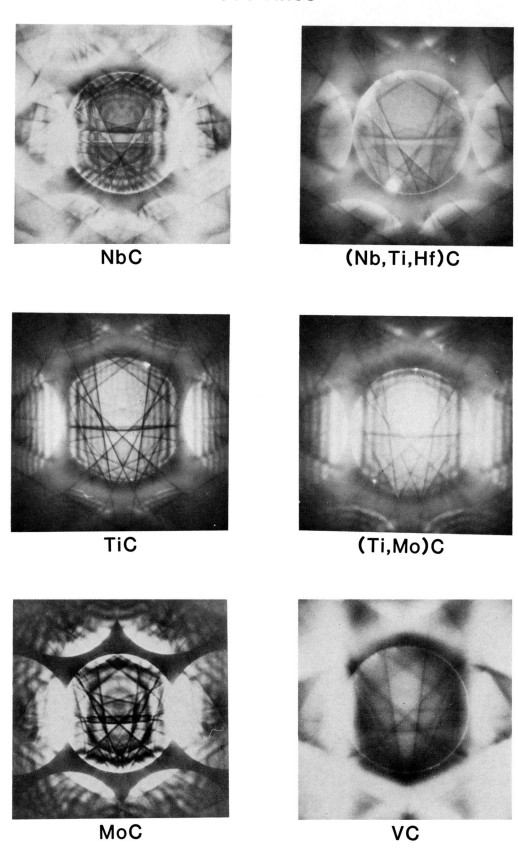

NbC

(Nb,Ti,Hf)C

TiC

(Ti,Mo)C

MoC

VC

57

Figure 10.5 (left) The ⟨100⟩ CBD pattern from NbC contains a series of concentric rings in the bright field disc similar to figure 10.3. Once again the material is close to a critical voltage at 100 kV. The faint HOLZ lines crossing these concentric fringes indicate an exceptionally perfect lattice.

Figure 10.6 (below) ⟨113⟩ axis of MnS. The structure factors (F_{hkl}) for MnS are determined by whether the Mn and the S atoms scatter in or out of phase with each other. With h, k, l all even, as in the SOLZ, F is proportional to $(f_{Mn} + f_S)$ and with h, k, l all odd, as in the FOLZ, F is proportional to $(f_{Mn} - f_S)$. As the HOLZ reflections behave in an approximately kinematic fashion, their intensity is roughly proportional to $|F|^2$, and hence the FOLZ ring is much weaker than the SOLZ ring in the figure. The increase in structure factor on moving from the FOLZ to the SOLZ more than compensates for the large increase in Debye–Waller factor.

Phase name: M_7X_3
Lattice type: orthorhombic/pseudohexagonal cf Mn_7C_3
Point group: *mmm*
Space group: *Pmna*
Lattice parameters: $a = 0.452$ nm, $b = 0.699$ nm,
$c = 1.211$ nm [1]

FORMULA AND COMPOSITIONAL VARIABILITY

In 316 steel the phase formed by reaction with the carbon extraction replica under electron beam irradiation and its composition was directly related to the phase which reacted to form it. The composition was approximately $(Cr,Fe)_7X_3$. Hyde *et al* [2] described Fe_7C_3 as having the hexagonal Ru_7B_3 structure but Mn_7C_3 has a pseudo-hexagonal orthorhombic structure [1]. The diffraction patterns we obtained were consistent with the orthorhombic cell.

MORPHOLOGY

The carbide formed by reaction with the replica was in the form of thin, faulted sheets. In 718 alloy there was no distinctive morphology, although the particles often had a complex structure of planar faults on more than one set of planes. These faults produce marked streaking in the diffraction patterns.

OCCURRENCE

$(Cr,Fe)_7C_3$ can occur in alloy steels [3]. The nucleation sequence is uncertain. It either forms directly from ferrite or else from cementite within the ferrite. It is not generally observed in the 300 series steels. However, M_7X_3 can be produced accidentally by beam heating the boride M_2B, manganese sulphide or the intermetallic σ or Laves phases on carbon extraction replicas [4]. In superalloys M_7X_3 has been identified in the region of electron beam welds. In this case the main constituents were Cr and Fe with a trace of Ni.

REFERENCES

1 Rouault A, Herpin P and Fruchart R 1970 *Ann. Chim.* **5** 461
2 Hyde B G, Bagshaw A W, Anderson S and O'Keefe M 1974 *Ann. Rev. Mater. Sci.* **4** 43
3 Dyson D J and Andrews K W 1969 *J. Iron Steel Inst.* **207** 208
4 Cooke K E 1977 *Developments in Electron Microscopy and Analysis 1977* (Inst. Phys. Conf. Ser. 36) p 431

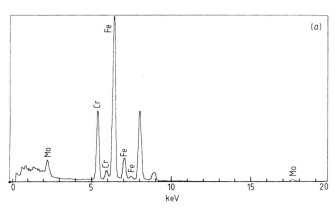

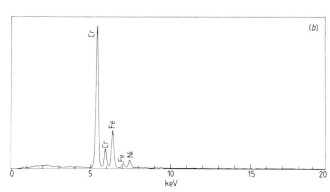

Figure 11.1 EDX spectra from M_7X_3 particles on (*a*) a CER of 316 steel and (*b*) a CER of an electron beam weld in a specimen of IN 718. Note in (*a*) the spectrum is almost the same as that of the σ phase from which it formed.

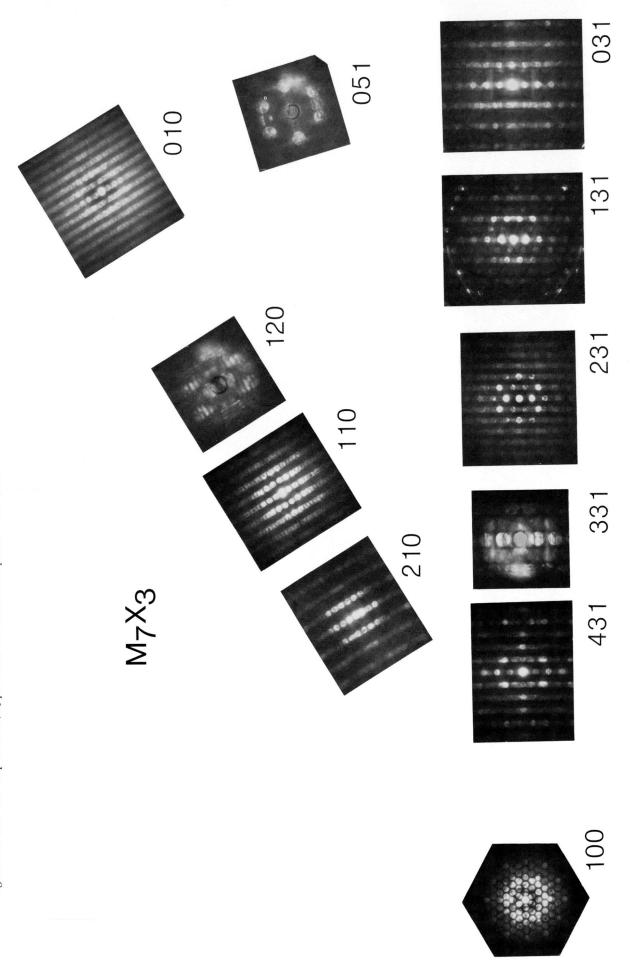

Figure 11.2 CBED ZAP map from M_7X_3 produced on extraction replicas of 316 steel.

Phase name: M_2X, chromium subnitride Cr_2N, dimolybdenum carbide Mo_2C

Lattice type and lattice parameters: These phases are members of a family of compounds in which the metal atoms form a hexagonal close packed lattice, with $c/a \sim \sqrt{(8/3)}$. The lighter atoms occupy approximately half of the octahedral interstitial sites. Ordering of the light atoms gives rise to unit cells which are larger than the metal sublattice unit cell and may be hexagonal or orthorhombic. In Cr_2N a larger hexagonal cell has been observed [1] with space group $P\bar{3}1m$ with $a_s = \sqrt{3}a_{hex}$ and $c_s = c_{hex}$, where s and hex refer to the superlattice and the hcp lattice respectively. Nitrogen is a weaker scatterer than chromium so that special care is required in determining the space and point groups of the ordered phase.

Mo_2C is related to Cr_2N in that it has a hcp lattice with an interstitial element which may or may not be ordered. The disordered cell has the same dimensions as the metal sublattice, i.e. $a_{hex} = 0.3002$ nm, $c_{hex} = 0.4724$ nm. In this the carbon atoms fill the two distinct interstitial sites randomly with occupancy of one half (β-Mo_2C [2]). A second arrangement, with an ordered arrangement of carbon atoms, gives an orthorhombic superlattice (ζ-Fe_2N type) with $a = 0.472$ nm($= c_{hex}$), $b = 0.600$ nm ($= 2a_{hex}$) and $c = 0.519$ nm($= \sqrt{3}a_{hex}$). This is known as α-Mo_2C [3]. A third variant exists with the carbon atoms sitting in an ordered hexagonal cell having $a = \sqrt{3}a_{hex}$ and $c = c_{hex}$ (for further information see [4]).

Point group: Cr_2N: $6/mmm$, α-Mo_2C: mmm, β-Mo_2C: $6/mmm$

Space group: Cr_2N: $P6_3/mmc$, α-Mo_2C: $Pbcn$, β-Mo_2C: $P6_3/mmc$

FORMULA AND COMPOSITIONAL VARIABILITY

Although there are only two explicit examples mentioned above, there are many similar compounds with the general formula M_2X. Examples are known with:

M as V, Cr, Fe, Mo, Nb, Ta and

X as C or N.

The c/a ratio of the hexagonal forms is markedly dependent on the M:C ratio [5]. Boron has not been included in the list of X elements because the metal atoms of M_2B do not sit on a simple hcp lattice (see § 14).

MORPHOLOGY AND OCCURRENCE

Cr_2N occurs in 316 steel after aging in air, particularly in the surface layers of air-aged samples. Atmospheric nitrogen evidently diffuses into the steel and reacts with chromium in the matrix. It is frequently found in close proximity to precipitates of M_6X and the formation of the two phases may be to some extent interdependent [5]. Regions of high Cr_2N concentration tend to have a low density of $M_{23}X_6$, the nitrogen replacing the carbon. In 316 steel Cr_2N usually occurs as thin, sub-micron length platelets which often exhibit extinction contours. It can form in needles in ferritic steels.

Mo_2C has not been observed in our 316 samples. It was studied as grain boundary precipitates in molybdenum and was subsequently produced by direct reaction from the elements. The effects of the orthorhombic arrangement of carbon atoms are easily missed because of the weak scattering power in comparison with the molybdenum. Nevertheless, the orthorhombic symmetry is visible with care. Another indication of the ordering of the carbon atoms in this phase is the appearance of very weak linear features in Mo_2C precipitates.

REFERENCES

1 Eriksson S 1934 *Jernkontorets Ann.* **118** 530

2 Rudy E, St Windisch, Stosick A J and Hoffman J R 1967 *Met. Trans.* A **239** 1247

3 Parthe E and Sadagopan V 1963 *Acta Crystallogr.* **16** 202

4 Andrews K W and Hughes H 1959 *J. Iron Steel Inst.* **194** 304

5 Thier H, Baumel H A and Schmidtman E 1969 *Arch. Eisen.* **40** 333

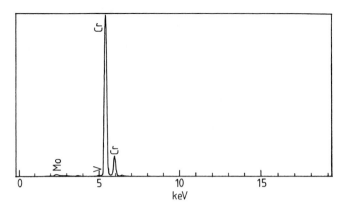

Figure 12.1 An EDX spectrum from a precipitate of Cr₂N on a CER of 316 steel.

Figure 12.2 (*right*) CBED ZAP map from chromium subnitride. The map is based on a hexagonal unit rather than a trigonal one, because it has not been possible to confirm the point group with the patterns so far recorded.

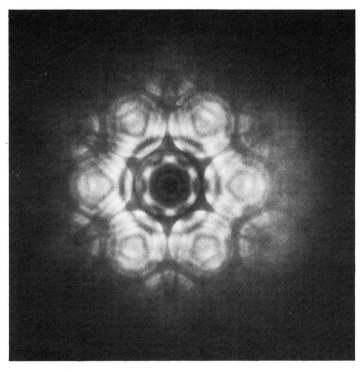

Figure 12.3 The [0001] axis of Cr₂N at long camera length. No HOLZ detail is visible, but the projection symmetry is clearly *6mm*.

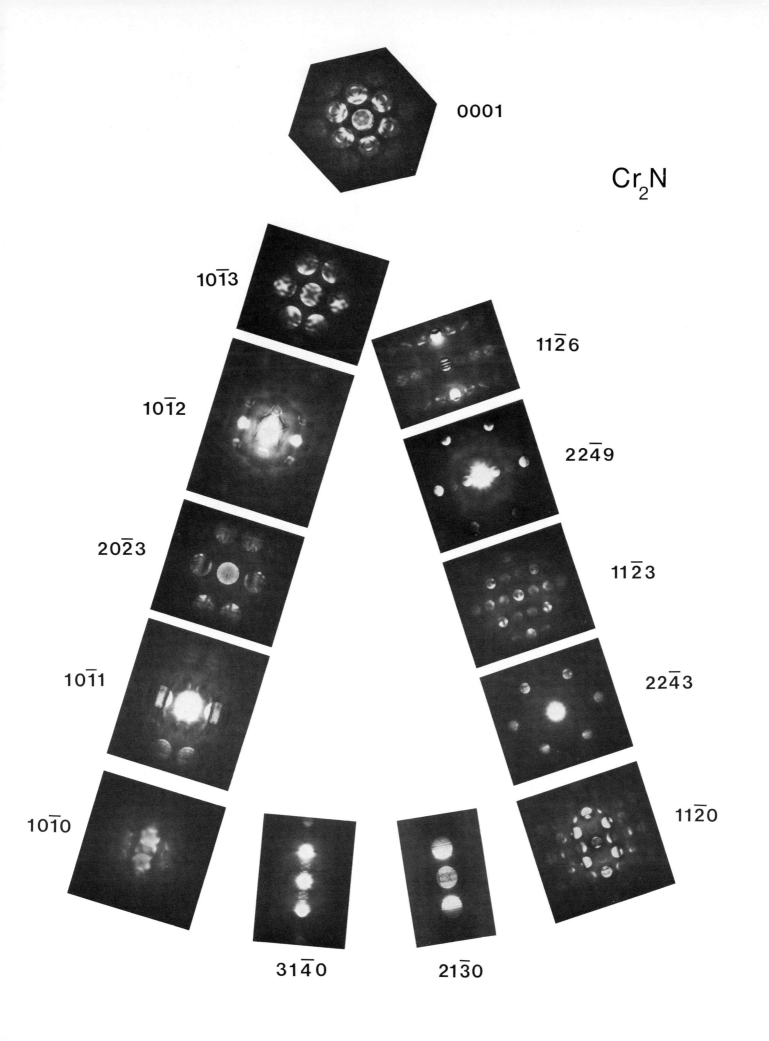

0001

Cr$_2$N

10$\bar{1}$3

10$\bar{1}$2

20$\bar{2}$3

10$\bar{1}$1

10$\bar{1}$0

11$\bar{2}$6

22$\bar{4}$9

11$\bar{2}$3

22$\bar{4}$3

11$\bar{2}$0

31$\bar{4}$0

21$\bar{3}$0

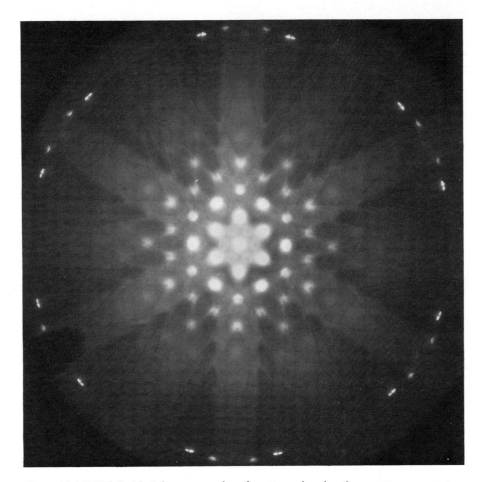

Figure 12.4 [0001] Cr$_2$N. A low camera length pattern showing *6mm* pattern symmetry. The 1$\bar{1}$00 reflections are visible because of the ordering of the nitrogen atoms. The intensity of these reflections is low, due to the weak scattering power of nitrogen as compared with the chromium atoms.

Figure 12.5 [010]/[0001] convergent beam pattern from a fragment of Mo$_2$C produced from the direct reaction of graphite and molybdenum. The HOLZ line detail in the central disc reveals *2mm* symmetry as expected for the orthorhombic form (β-Mo$_2$C). Note however the near-hexagonal symmetry of the pattern which is dominated by the scattering from the molybdenum atoms.

Phase name: M_3B_2
Lattice type: primitive tetragonal, uranium silicide (U_3Si_2) type [1]
Point group: $4/mmm$
Space group: $P4/mbm$
Lattice parameters: $a = 0.310$ nm, $c = 0.570$ nm in 316 steel
$a = 0.313$ nm, $c = 0.578$ nm in IN 718

M_3B_2

FORMULA AND COMPOSITIONAL VARIABILITY

This phase is usually of the form Mo_3B_2 with significant substitution of Mo by elements such as Cr, Fe, Ni, Ti, Co, Nb and V, e.g.

in 316 steel $(Mo,Cr)_2(Fe,Cr)B_2$
in superalloys $(Mo,Ti)_2(Cr,Ni,Co)B_2$ (the most common case)
or $(Mo,Ti)(Cr,Ni,Co)_2B_2$.

OTHER FEATURES AND COMMENTS

This phase does not seem to form in very high boron alloys, as M_2B (see §14) tends to form instead.

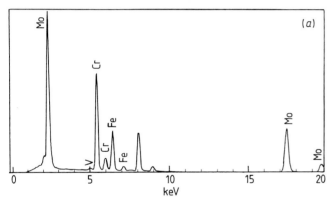

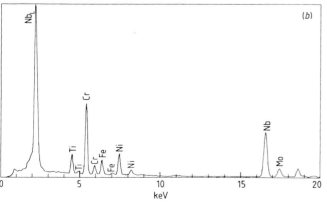

OCCURRENCE AND MORPHOLOGY

In 316 steel, with high boron levels, this phase is found as rounded particles which range from 0.5–3 μm in length. It is metallurgically inactive and does not dissolve during solution treatment.

In superalloys, although the boron content is beneficial to the creep rupture properties (as in many steels), the presence of borides may be detrimental; for example, they may reduce the amount of boron that is available to benefit the mechanical properties. Also, boron is thought to contribute to liquation-cracking in the heat-affected zone of electron-beam welds. In these welds the boride occurs as small sub-micron particles. Grain boundary borides occur as large blocky particles, often several microns in length.

REFERENCE

1 Pearson W B 1972 *The Crystal Chemistry and Physics of Metals and Alloys* (New York: Wiley)

Figure 13.1 Characteristic EDX spectra from the boride M_3B_2 found in (*a*) high boron 316 steel and (*b*) and (*c*) electron beam welds in IN 718. This boride also occurs in IN 901 and Astroloy. In both alloys the spectra are similar to (*b*) and (*c*), depending on whether the alloy contains Nb, Mo or both elements.

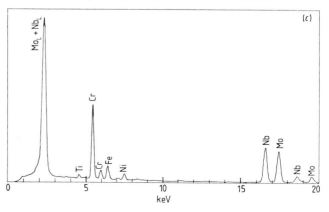

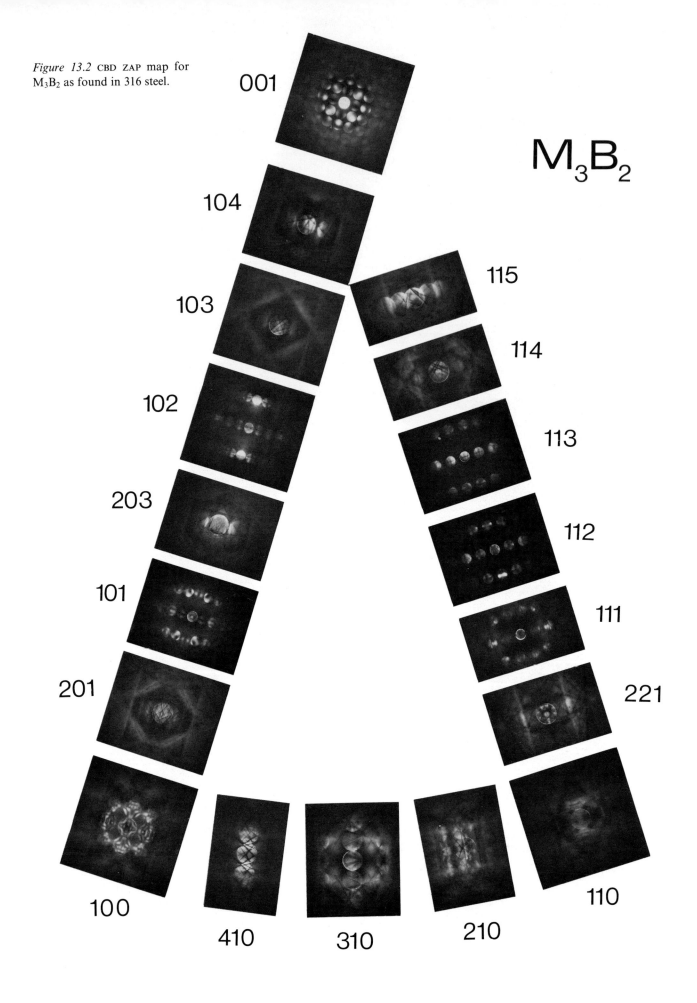

Figure 13.2 CBD ZAP map for M₃B₂ as found in 316 steel.

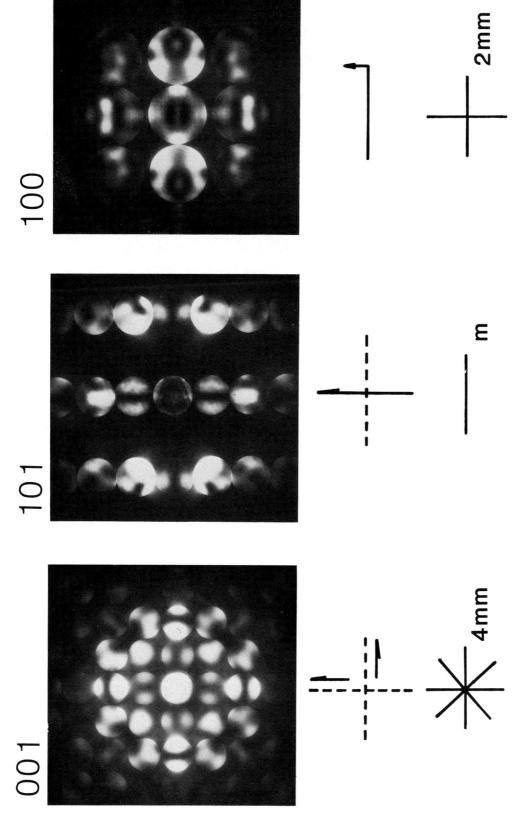

M₃B₂

100

101

001

2mm

m

4mm

Figure 13.3 Three CBD patterns observed when tilting about the [010] axis. On the left the beam is parallel to the tetrad axis and the pattern symmetry is *4mm*. The dark bars in the ±100 and ±010 reflections are due to both screw axes and glide planes. Tilting to the centre pattern, the [101] axis, the dark bars in the alternate horizontal and vertical directions are due to the {010} glide planes and ⟨010⟩ screw diad respectively. HOLZ lines can be discerned in the central disc and the pattern symmetry is *m*. Further tilting leads to the [100] axis pattern which is printed to the same scale as the preceding two patterns, although a larger condenser aperture has been used. The HOLZ detail is faint but a pattern symmetry of *2mm* can be seen. The significant space and point group elements have been sketched beneath each axis. Patterns recorded at 120 kV.

67

M₂B

Phase name: M₂B, 'iron chromium boride'
Lattice type: orthorhombic, pseudohexagonal, based on Mn₄B structure [1]
Point group: *mmm*
Space group: *Fddd*
Lattice parameters: $a = 1.457$ nm, $b = 0.732$ nm, $c = 0.422$ nm [2,3]

OCCURRENCE

This phase usually occurs in high boron steels. It is reported to be more stable than M₃B₂ [4]. In this work it was found in a steel with a composition between that of 316 and 304 and a boron content much higher than is usual in either alloy [5].

OTHER FEATURES AND COMMENTS

Occasionally, particles of this boride have been observed to react with the carbon support film when illuminated with an electron beam.

MORPHOLOGY

This phase occurred as large irregular blocks at grain boundaries and triple points. It contained extensive linear disorder which was analysed as coming mostly from rotation twins. As a result of this fine scale twinning the patterns generally indicated a hexagonal phase, and hence it could be easily confused with other heavily faulted hexagonal or pseudohexagonal phases such as Laves, M₇X₃ and SiC.

REFERENCES

1 Kiessling R 1950 *Acta Chem. Scand.* **4** 146
2 Brown B E and Beerntsen D J 1964 *Acta Crystallogr.* **17** 448
3 Guttmann-Senicourt D and Strudel J L 1974 *Rev. Met.* **71** 329
4 Goldschmidt H J 1967 *Interstitial Alloys* (London: Butterworth)
5 Stoter L P 1981 *J. Mater. Sci.* **16** 1039

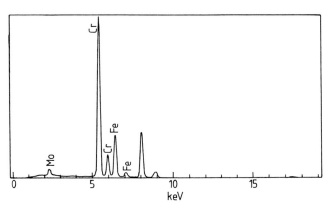

Figure 14.1 Typical EDX spectrum of M₂B in 316 steel. Note that M₂₃X₆ can have a similar spectrum.

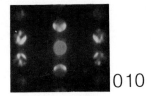 010

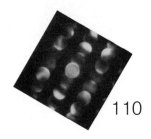 110

Figure 14.2 CBED ZAP map for the boride M_2B.

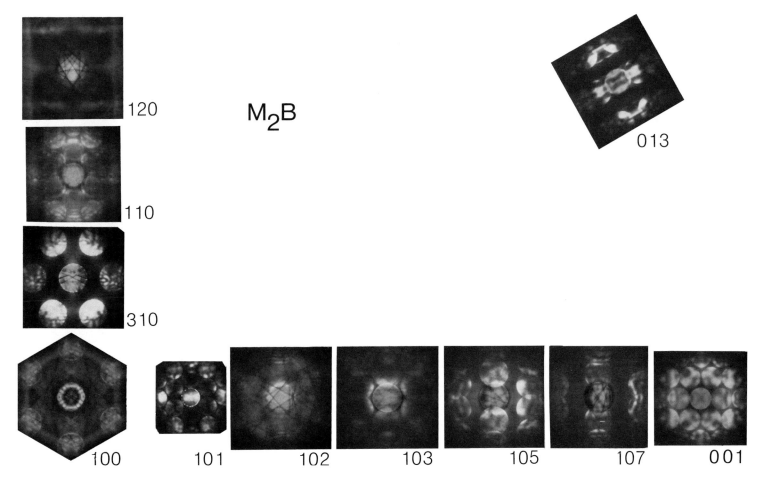
120
110
310
M₂B
013
100   101   102   103   105   107   001

69

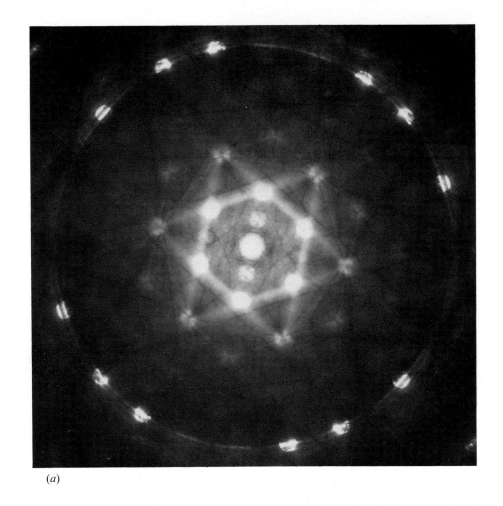

(a)

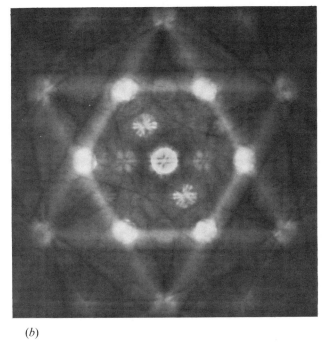

(b)

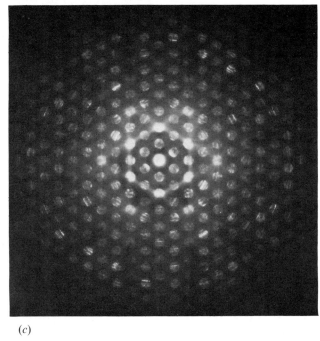

(c)

70

Figure 14.3 (left) (a) A low camera length [100] ZAP of M₂B equivalent to the [0001] hexagonal axis. Note particularly that the crystallite from which this pattern was obtained did not contain twins, so that only one direction linking the six strong reflections to the direct beam is subdivided (by weak 020 reflections). Both the bright field and the whole pattern symmetries are *2mm*. Two further [100] M₂B ZAPs illustrate the effects of faulting when the electron probe covers two or more twin orientations. *(b)* This pattern was recorded from a region with two twin orientations rotated $\pm\pi/3$ or $\pm 2\pi/3$ about [100]. The twinning has given rise to a second set of weak inner reflections. *(c)* A low camera length pattern from a heavily twinned crystal. There are three sets of weak inner reflections in the 020 positions and the zero-layer discs extend out to large angle, implying streaking along [100]. Selected area diffraction patterns would probably be interpreted as coming from a hexagonal crystal.

Figure 14.4 [100] M₂B at long camera length showing the bright field and whole pattern symmetry of *2mm*. The 020 reflections are almost invisible for this rather thick crystal.

$M_4C_2S_2$

Phase name: titanium carbosulphide. More general form $M_4C_2S_2$ (see below)
Lattice type: hexagonal
Point group: $6/mmm$
Space group: $P6_3/mmc$
Lattice parameters: $Ti_4C_2S_2$ $a = 0.321$ nm, $c = 1.120$ nm, $c/a = 3.49$ [1]
$Zr_4C_2S_2$ $a = 0.340$ nm, $c = 1.211$ nm, $c/a = 3.50$ [1]

FORMULA AND COMPOSITIONAL VARIABILITY

Although based on titanium carbosulphide there can be other elements present, e.g. niobium and zirconium.

OCCURRENCE

The phase has been found in superalloys, e.g. Astroloy, Pyromet, 713LC, IN 718, IN 10 and D979 [2,3,4,5]. It has also been observed in pig iron [6].

OTHER FEATURES AND COMMENTS

Identification of this phase may be complicated by the existence of two other titanium sulphides with similar space groups and lattice constants. These are
Ti_3S_4 $a = 0.344$ nm, $c = 1.145$ nm, $P6_3mc$
Ti_2S_3 $a = 0.343$ nm, $c = 1.142$ nm, $P6_3mc$.

MORPHOLOGY

In most cases it occurred as plates or laths with the faces parallel to the basal plane. In Astroloy this precipitate was found on grain boundaries in association with MX, borides and ZrO_2 (see §§ 10, 13, 14 and 23 respectively). In Astroloy the phase occurred as 0.05–0.2 μm sized particles or plates 0.5 μm wide.

REFERENCES

1 Johnson V and Jeitschko W 1972 *J. Solid State Chem.* **4** 123
2 Wallace W, Holt R T and Terada T 1973 *Metallography* **6** 511
3 Whelan E P and Grzedzielski M S 1974 *Met. Tech.* **1** 186
4 Donachie H J and Kriege O H 1972 *J. Mater.* **9** 269
5 Vincent R and Bielicki T A 1982 Applications of CBD to Nickel Superalloys in *Electron Microscopy and Analysis 1981* (Inst. Phys. Conf. Ser. 61) p 491
6 Kudielka H and Rohde H 1960 *Z. Krist.* **114** 441

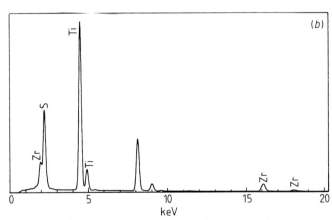

Figure 15.1 Examples of EDX spectra from $M_4C_2S_2$ particles found in superalloys. (*a*) This spectrum was obtained from a carbosulphide particle near an electron beam weld in a sample of Inconel 718. To avoid confusion with the mixed (Nb,Ti)C precipitate from the same alloy note the strong sulphur $K_{\alpha+\beta}$ peak at 2.3 keV which almost overlaps the Nb $L_{\alpha+\beta}$ peak. (*b*) This spectrum was generated from a carbosulphide discovered in a sample of Astroloy.

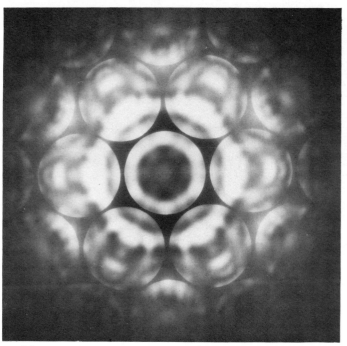

(a)

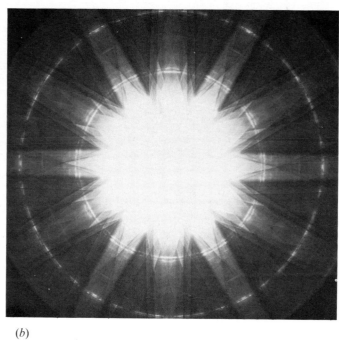

(b)

Figure 15.2 An [0001] CBED pattern from a carbosulphide precipitate in Inconel 718. (*a*) Long camera length. The projection symmetry is *6mm*. Pattern recorded at 120 kV. (*b*) Low camera length showing two HOLZ rings, the first and the third. The second ring is absent because it is composed of kinematically weak reflections. The pattern symmetry deduced from the structure of the HOLZ rings is *6mm*.

M₃P

Phase name: M₃P (chromium phosphide)
Lattice type: tetragonal
Point group: $\bar{4}$
Space group: $I\bar{4}$
Lattice parameters: $a = 0.911$ nm, $c = 0.460$ nm [1]

FORMULA AND COMPOSITIONAL VARIABILITY

The structure is based on Cr_3P but generally contains both chromium and iron. As or B may also be present as substitutions for P [1].

MORPHOLOGY

In neutron-irradiated steels this phase was found as thin sheets or plates (typically several μm long). It nucleated preferentially on chromium subnitride particles.

OCCURRENCE

Phosphorus is a very undesirable element, having long been associated with poor mechanical properties. Its presence in steels and other commercial alloys is usually minimised and, although phosphorus segregation remains a potential problem, phosphides are not generally observed. Chromium phosphide has, however, been identified in neutron-irradiated steels and it would probably occur in any region where there is strong phosphorus segregation.

REFERENCE

1 Rundqvist S 1962 *Acta Chem. Scand.* **16** 1

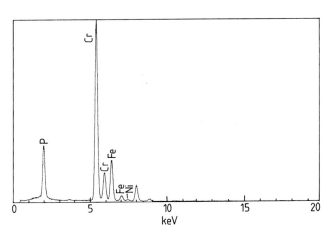

Figure 16.1 EDX spectrum from a M₃P particle found in a neutron-irradiated ferritic steel.

Figure 16.2 The [001] zone axis of M₃P. It is unusual in showing 4-fold projection symmetry without mirror lines.

MM′P

Phase name: MM′P
Lattice type: orthorhombic
Point group: *mmm*
Space group: *Pnma*
Lattice parameters: $a = 0.611$ nm, $b = 0.350$ nm, $c = 0.709$ nm [1]

FORMULA AND COMPOSITIONAL VARIABILITY

This phase has two essentially distinct metal atom sites, similar to M_3B_2.

MORPHOLOGY

MM′P occurred as either rounded or tabular particles with a size range of 0.1–0.5 μm. The precipitates frequently contained internal grain boundaries but no fault planes were seen (cf Laves precipitates).

OCCURRENCE

MM′P is not a commonly encountered phase. The micrographs presented in this section were recorded from small particles discovered in the heat-affected zone of an electron-beam weld in a sample of Inconel 718.

REFERENCE

1 Rundqvist S and Nawapong P C 1966 *Acta Chem. Scand.* **20** 2250

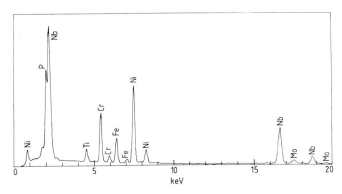

Figure 17.1 A typical EDX spectrum from the ternary phosphide found in IN 718. It is distinguished from the Laves phase in that alloy by the enhanced Nb K_α peak and the presence of a phosphorus $K_{\alpha+\beta}$ peak which almost overlaps the Nb $L_{\alpha+\beta}$ peak.

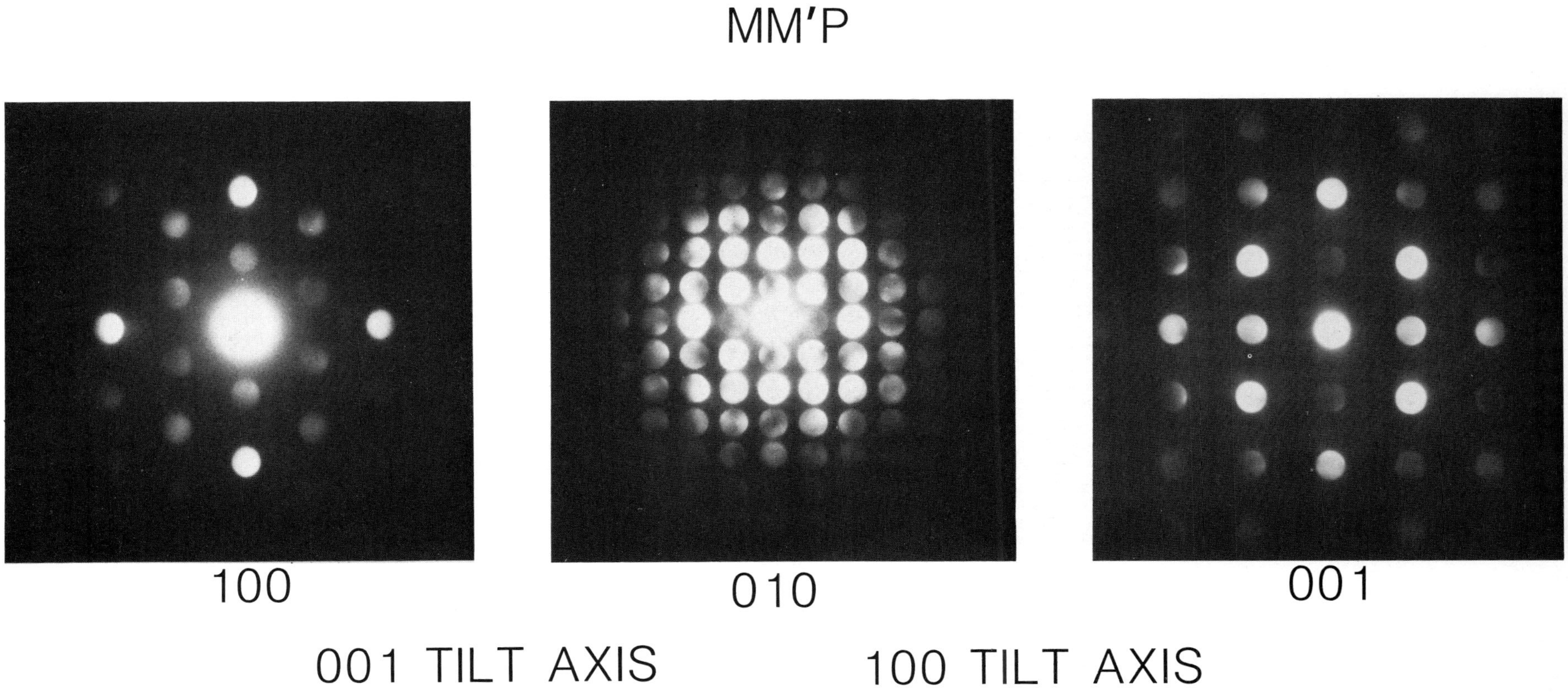

MM'P

001 TILT AXIS   100 TILT AXIS

100   010   001

Figure 17.2 CBED patterns from the three orthogonal axes of MM'P. Patterns recorded at 120 kV. (*a*) [100] axis. (*b*) [010] axis. It contains somewhat imperfect dark bars in the alternate reflections in the 100 and 001 systematic rows resulting from the glide planes on (100) and (001). This pattern is related to (*a*) by a common vertical 00*l* systematic row. (*c*) [001] axis. This pattern is related to (*b*) by a common horizontal *h*00 systematic row; however, the 100 reflections are absent because of the lack of double diffraction paths. The 010 reflections appear very weakly at this axis because of the existence of weak double diffraction routes.

TiO$_2$ rutile and anatase

Phase name: titanium dioxide (rutile and anatase)
Lattice type: both are tetragonal
Point group: both are 4/*mmm*
Space group: rutile *P*4$_2$/*mnm*, anatase *I*4$_1$/*amd*
Lattice parameters: rutile $a = 0.460$ nm, $c = 0.296$ nm
anatase $a = 0.380$ nm, $c = 0.950$ nm

FORMULA AND COMPOSITIONAL VARIABILITY

A study of the deoxidisation of iron by titanium has shown that there are a considerable number of possible oxides of Ti and Ti with some Fe substitution [1]. However, we have only come across the two forms listed above (this leaves Brookite as the only other common form not encountered).

MORPHOLOGY

In 316 steel these phases occurred as small (1 μm or less in size) rounded particles. Some were large enough to show extinction contours. The particles were often found in clusters.

OCCURRENCE

Titanium, as well as being a common alloying element in steels, is also present in some refractories and welding electrodes. The free energy of formation for TiO$_2$ is low [1]. Hence this oxide may be formed in manufacture or may be introduced as an impurity from the furnace.

REFERENCE

1 Kiessling R and Lange N 1978 *Non-metallic Inclusions in Steel* (London: The Metals Society)

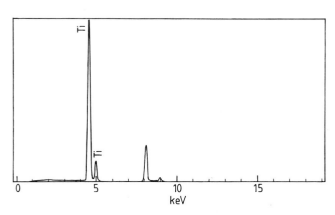

Figure 18.1 EDX spectrum of TiO$_2$ as found in 316 steel (typical of rutile or anatase). Only Ti K$_\alpha$ and K$_\beta$ peaks are visible.

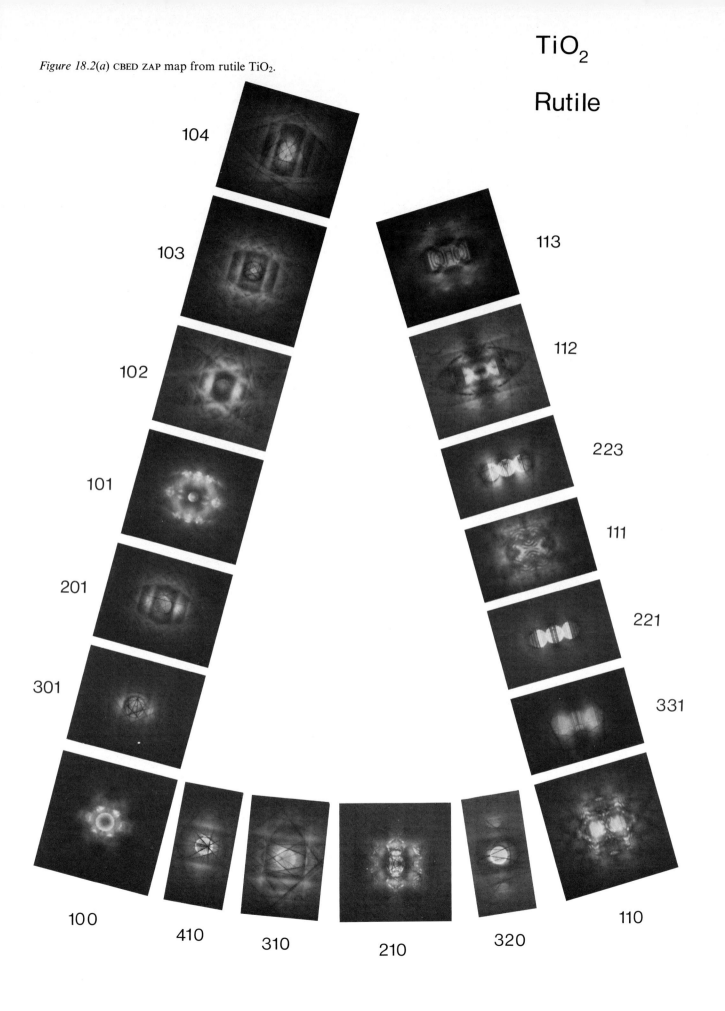

TiO$_2$

Rutile

Figure 18.2(a) CBED ZAP map from rutile TiO$_2$.

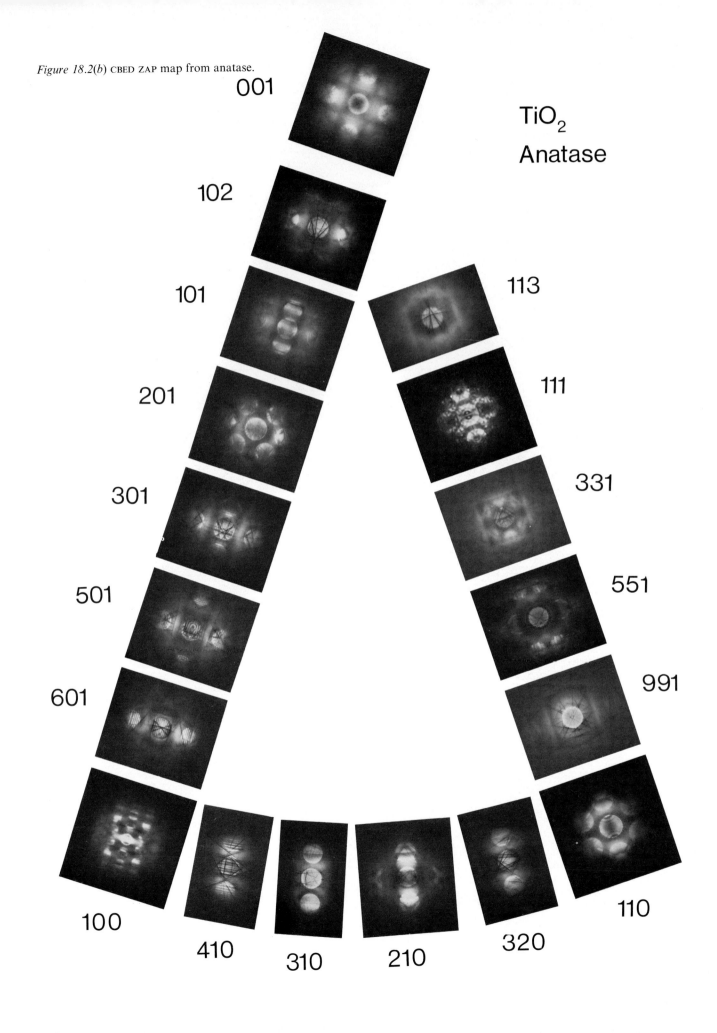

Figure 18.2(b) CBED ZAP map from anatase.

001

102

101

201

301

501

601

100

410

310

210

320

110

113

111

331

551

991

TiO_2
Anatase

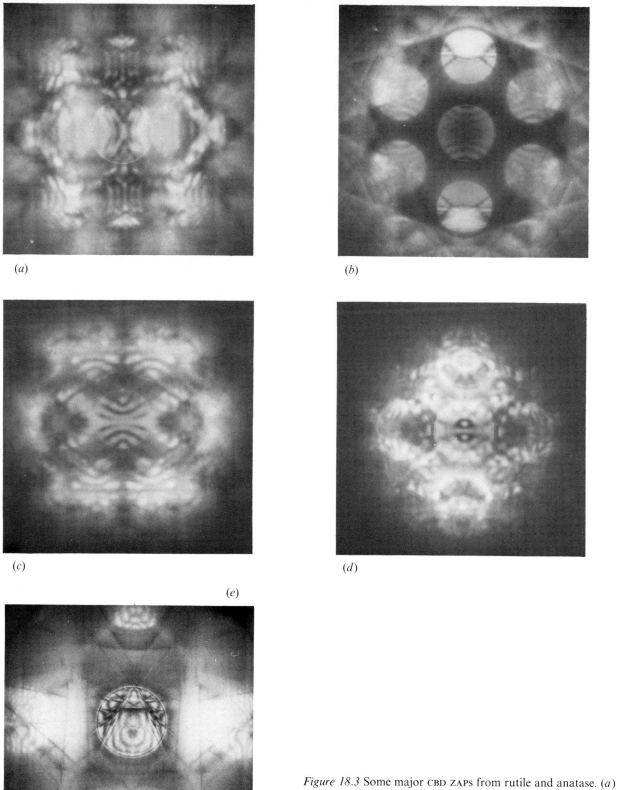

(a)

(b)

(c)

(d)

(e)

Figure 18.3 Some major CBD ZAPs from rutile and anatase. (*a*) ⟨110⟩ rutile. (*b*) ⟨110⟩ anatase. (*c*) ⟨111⟩ rutile. (*d*) ⟨111⟩ anatase. These axes are sufficiently distinctive to allow the phase to be identified. (*e*) ⟨051⟩ rutile. This pattern is presented because it contains interesting HOLZ detail and a simple fringe system in the direct beam.

α-alumina

Phase name: α-alumina (corundum)
Lattice type: trigonal, Cr_2O_3-type
Point group: $\bar{3}m$
Space group: $R\bar{3}c$
Lattice parameters: $a = 0.4759$ nm, $c = 1.2991$ nm, $c/a = 2.73$

FORMULA AND COMPOSITIONAL VARIABILITY

Although the EDX spectra cannot reveal the oxygen, they do confirm that the metal present is almost entirely aluminium with only traces of other elements.

OCCURRENCE

This phase is occasionally found on carbon extraction replicas of steels and alloys. It is a well known inclusion [1] and can occur as straight Al_2O_3 or in a multi-phase glassy inclusion with other oxides. It can also be introduced in the mechanical polishing necessary for the preparation of samples. This can happen even in the most careful preparations.

MORPHOLOGY

Large laths or idiomorphic plates have been observed in various steels, often several tens of μm long in low-grade alloys. In 316 steel it has been observed as rounded particles with approximately the same size as the $M_{23}X_6$ precipitates ($< 1\ \mu$m).

REFERENCE

1 Kiessling R and Lange N 1978 *Non-metallic Inclusions in Steel* (London: The Metals Society)

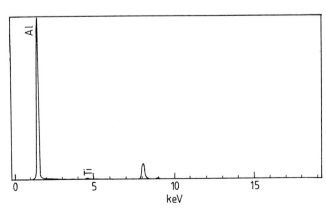

Figure 19.1 EDX spectrum from a piece of α-alumina discovered in a specimen of 316 steel. Note that there is a trace of titanium present.

Alpha Alumina [0001] zone axis pattern

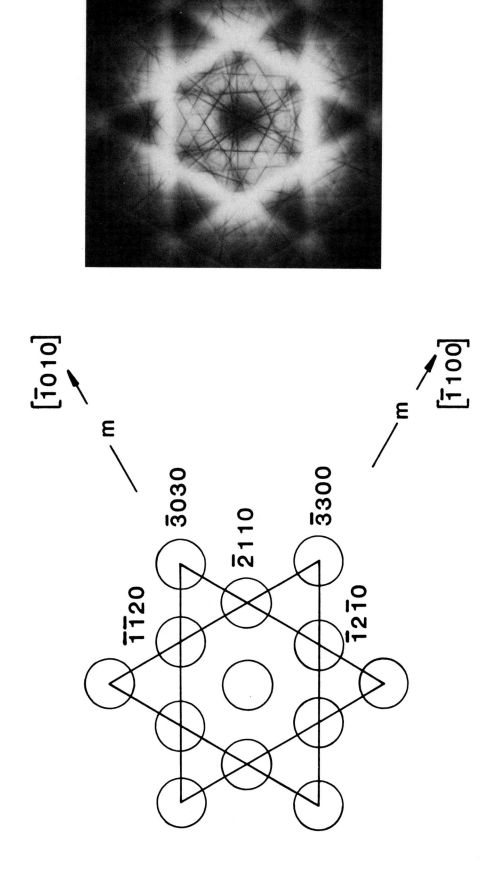

Figure 19.2 Although α-alumina is not hexagonal, it is most convenient to index diffraction patterns from trigonal crystals with Miller–Bravais indices. The diffraction pattern above is thus [0001], with $3m$ symmetry. The crystal was thick enough to show strong HOLZ contrast but the individual discs are almost totally invisible. The diagram accompanying the pattern is drawn to the same scale and is indexed. Note that the reflections nearest to the centre are not indexed as {10$\bar{1}$0} but {1$\bar{2}$10}. This is because the {10$\bar{1}$0} and the {20$\bar{2}$0} reflections are forbidden by the space group. The directions to the [$\bar{1}$100] and [$\bar{1}$010] axes are labelled.

83

Alpha Alumina (Corundum)

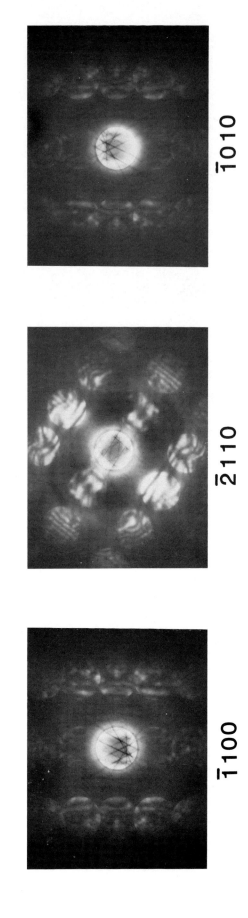

$\overline{1}100$

$\overline{2}110$

$\overline{1}010$

Figure 19.3 These three axes lie along the 0003 Kikuchi band which runs horizontally across the figure. Note that the axes are in the correct orientation for rotation about the vertical axis ([0001]). They clearly illustrate the trigonality of the crystal. Although the forms of the [$\overline{1}$100] and [$\overline{1}$010] patterns appear identical, the one is rotated by 180° with respect to the other, as required by the crystal symmetry. Also, the axis midway between them, [$\overline{2}$110], clearly lacks a mirror line normal to the 0003 Kikuchi band. The pattern has 2-fold symmetry.

Phase name: aluminium nitride, AlN
Lattice type: hexagonal
Point group: *6mm*
Space group: *P6₃mc*
Lattice parameters: $a = 0.310$ nm, $c = 0.500$ nm [1]

FORMULA AND COMPOSITIONAL VARIABILITY
The nitrogen content cannot be determined from the EDX spectrum but there is evidently some silicon and chromium in addition to the aluminium.

MORPHOLOGY
This phase was found as thin platelets that ranged in size from a few tenths of a micron to several microns.

OCCURRENCE
This phase was observed on CERs of 316 steel. It was found on replicas taken from the outer regions of specimens which had been aged and crept in air.

REFERENCE
1 Donnay J D H and Ondik H N (ed) 1973 *Crystal Data Determinative Tables, Vol 2—Inorganic Compounds* (US Dep. Comm., NBS and JCPDS)

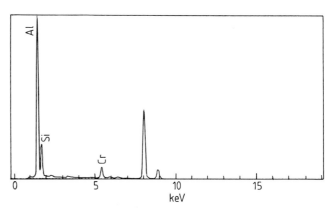

Figure 20.1 An EDX spectrum of an AlN particle extracted from a 316 steel sample. Note the Si and Cr content.

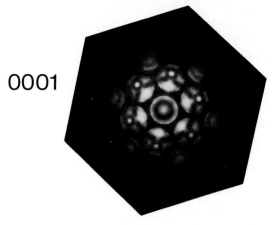

Aluminium Nitride

0001

Figure 20.2 CBED ZAP map of aluminium nitride. Particles of this phase were discovered in 316 steel. The crystals were very thin and therefore the HOLZ effects were weak: little detail is visible in the zero-layer discs.

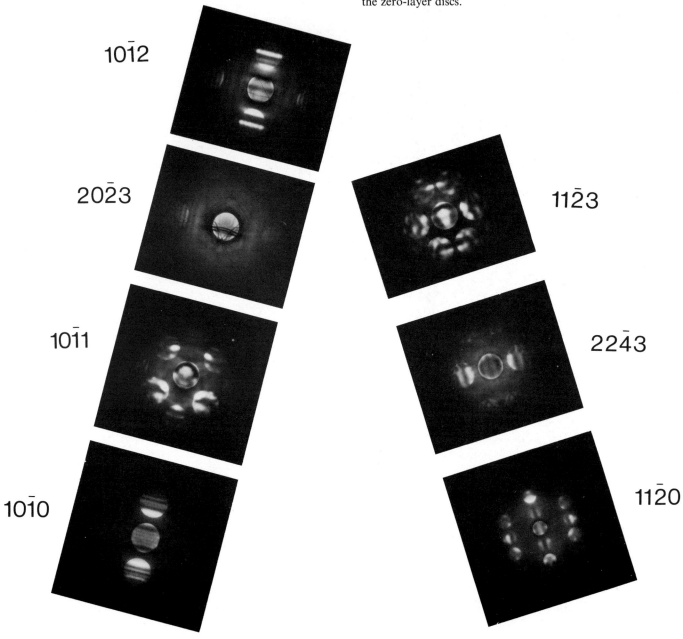

$10\bar{1}2$

$20\bar{2}3$

$11\bar{2}3$

$10\bar{1}1$

$22\bar{4}3$

$10\bar{1}0$

$11\bar{2}0$

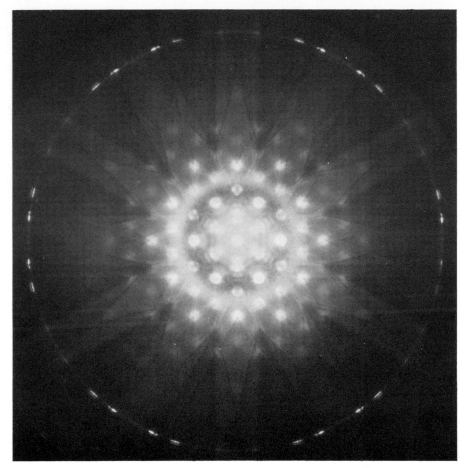

Figure 20.3 Low camera length [0001] diffraction pattern from AlN. The pattern symmetry is *6mm*.

Figure 20.4 CBD patterns from AlN, (*a*) ⟨1$\bar{1}$00⟩ axis and (*b*) ⟨11$\bar{2}$0⟩ axis. HOLZ lines are only faintly visible in the central disc of the ⟨1$\bar{1}$00⟩ pattern and the *2mm* symmetry is just discernible. The ⟨11$\bar{2}$0⟩ axis contains dark bars in the kinematically forbidden 0001 reflections as a result of the glide plane parallel to the beam and a perpendicular 2_1 screw axis.

(*a*)

(*b*)

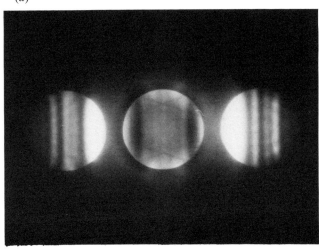

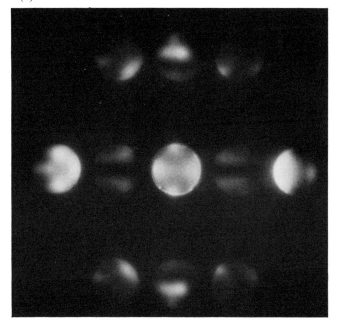

Spinel

Phase name: spinel
Lattice type: diamond cubic
Point group: $m3m$
Space group: $Fd3m$
Lattice parameters: $a = 0.851$ nm for the most common composition, $MnCr_2O_4$ (chromium galaxite) [1]

FORMULA AND COMPOSITIONAL VARIABILITY

Spinels have the general formula AB_2O_4 where A is characteristically a divalent metal such as Mg and B is a trivalent metal such as Al. In fact, there exists a great variety of spinels.

MORPHOLOGY

Spinels were usually found as large blocky particles often several μm in length. They were, presumably, inclusions in the steel unaffected by solution treatment.

REFERENCES

1 Kiessling R and Lange N 1978 *Non-metallic Inclusions in Steel* (London: The Metals Society)
2 Steeds J W and Evans N S 1980 A practical example of point and space group determination in convergent beam diffraction *Proc. 38th EMSA* (Claitors Publishing Division) p 188

OCCURRENCE

Many spinels have been discovered in steels [1]. In 316 steel both the classic spinel $MgO.Al_2O_3$ and chromium galaxite ($MnO.Cr_2O_3$) have been found. Particles with a mixture of Cr, Mn, Al and Ti have been identified with the spinel structure. They were possibly a mixed spinel of the form $(Mn,Ti)O.(Cr,Al)_2O_3$.

In Astroloy the spinel particles discovered contained Mg, Ti, and Al. These were probably of the form $(Mg,Ti)Al_2O_4$.

OTHER FEATURES AND COMMENTS

Although the space group is conventionally accepted as $Fd3m$, there have been a number of recent publications that have questioned this assignment. However, in the micrographs recorded for this text, there has been no indication of any deviation from the accepted space group [2].

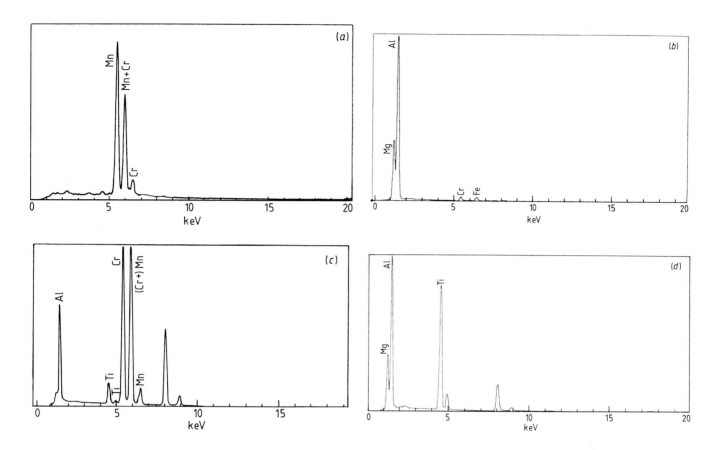

Figure 21.1 Three EDX spectra from spinel phases found in 316 steel and one spectrum from Astroloy. (*a*) Basically MnO.Cr$_2$O$_3$, (*b*) MgO.Al$_2$O$_3$, (*c*) a mixed version commonly observed and (*d*) (Mg,Ti)Al$_2$O$_4$ in Astroloy.

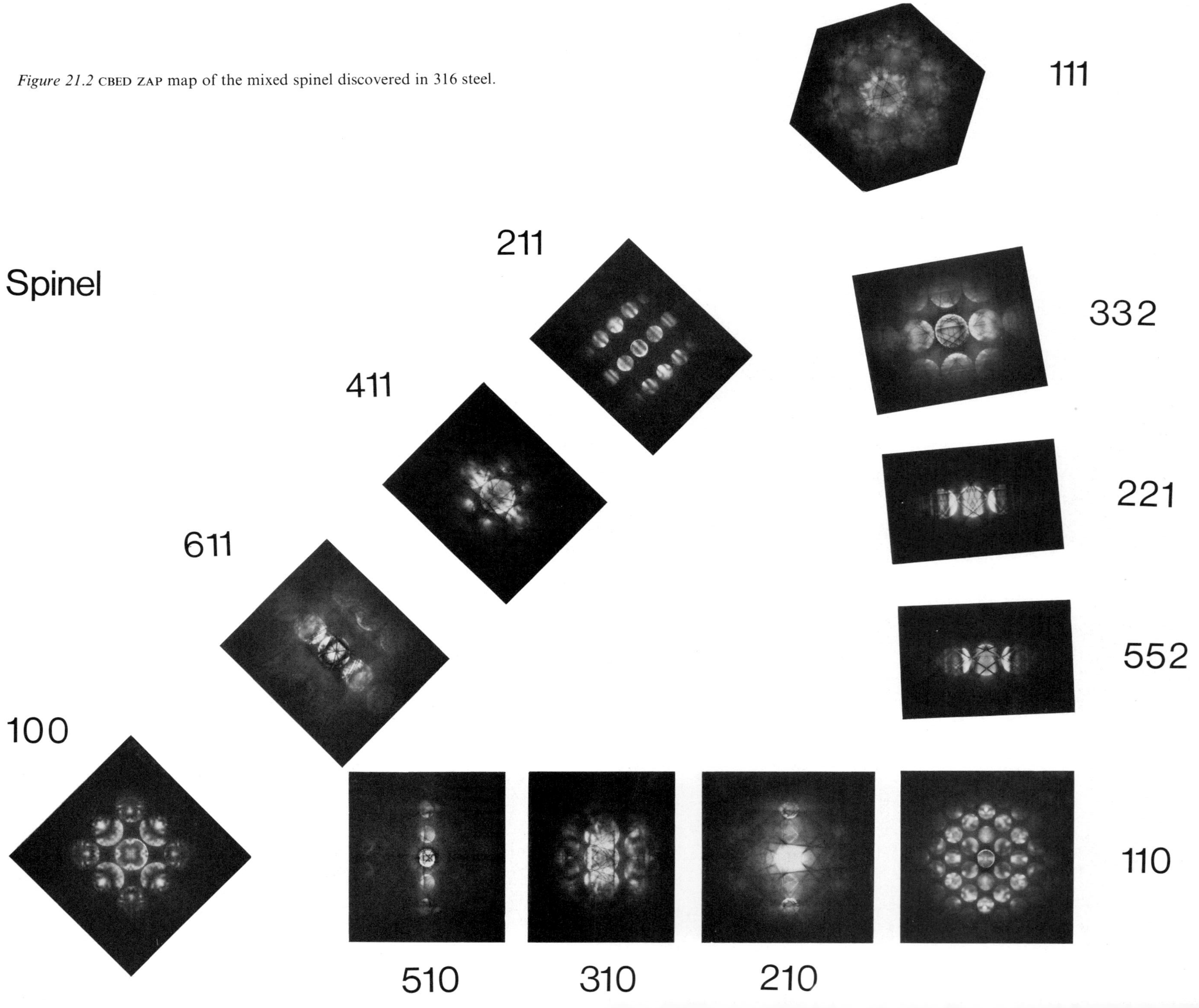

Figure 21.2 CBED ZAP map of the mixed spinel discovered in 316 steel.

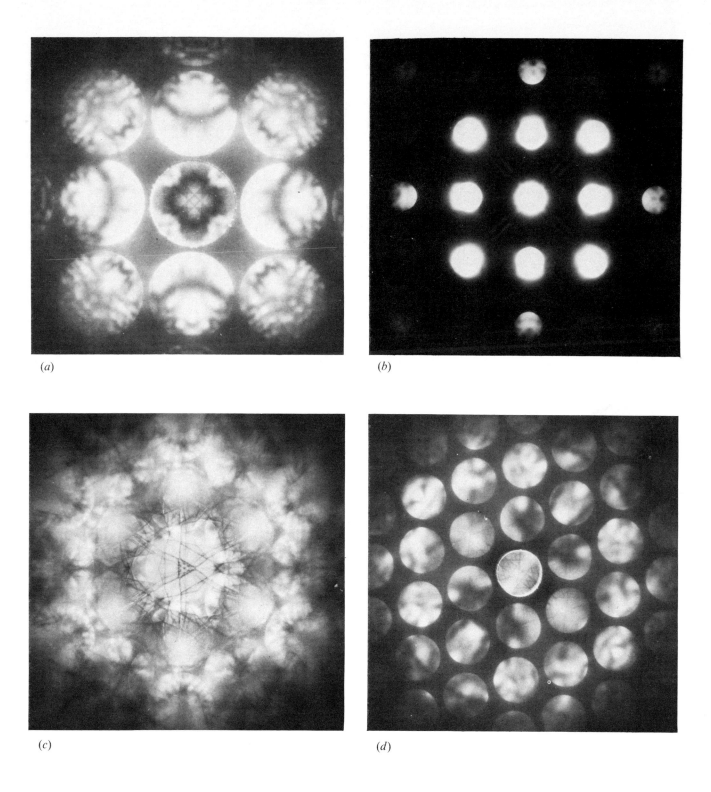

(a)

(b)

(c)

(d)

Figure 21.3 Four ZAPs from spinel particles found in 316 steel. (a) ⟨100⟩ axis recorded with a large condenser aperture. HOLZ lines are clearly visible in the zero order, the pattern symmetry is 4*mm*. The 002 discs are not visible. (b) ⟨100⟩ axis with a small condenser aperture. In this case no HOLZ line detail is visible in the central disc, but there are pairs of bright lines in the kinematically forbidden 002 discs as a result of double diffrac- tion via the FOLZ. (c) ⟨111⟩ axis. This pattern was recorded with a large condenser aperture. The diffuse scattering was so strong that the individual discs are barely discernible. 3*m* pattern symmetry is clearly visible. (d) ⟨110⟩ axis. Note that even with a small condenser aperture the 2*mm* whole pattern symmetry is well defined.

ZrO₂
HfO₂

Phase name: zirconium oxide ZrO₂ (zirconia)
hafnium oxide HfO₂ (hafnia)
Lattice type: monoclinic
Point group: $2/m$
Space group: $P2_1/a$
Lattice parameters: ZrO₂ $a = 0.5295$ nm, $b = 0.5172$ nm,
$c = 0.5116$ nm, $\beta = 99.18°$ [1]
HfO₂ $a = 0.5145$ nm, $b = 0.5208$ nm,
$c = 0.5311$ nm, $\beta = 99.23°$ [1]

FORMULA AND COMPOSITIONAL VARIABILITY

Zirconia in superalloys was found to contain traces of titanium and chromium (see EDX spectra). Hafnia also had trace impurities, of which zirconium was the most important, although there was sometimes a detectable amount of magnesium.

OTHER FEATURES AND COMMENTS

These oxides were found to be electron beam sensitive and therefore the diffraction patterns had to be recorded carefully to obtain reasonable detail. Beam damage of the particles caused a loss of HOLZ information.

OCCURRENCE AND MORPHOLOGY

Zirconium oxide was found in Astroloy, usually as discoid particles of the order of 0.05 μm in diameter. Zirconia, like TiO₂ in steels, has a low energy of formation and could form during manufacture of the alloys. It was generally found as small particles, too small to be confused with slags. Hafnium oxide has been found in powder alloys with hafnium additions. The particles were generally tabular and of the order of 0.2 μm in length. They could be confused with MX precipitates of similar dimensions (see § 10).

REFERENCE

1 Adam J and Roger M P 1959 *Acta Crystallogr.* **12** 951

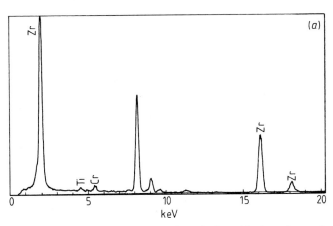
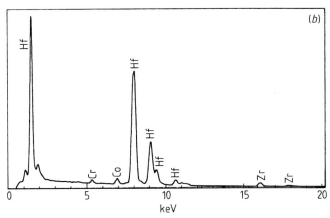

Figure 22.1 EDX spectra from (*a*) zirconia found in Astroloy and (*b*) hafnia found in a hafnium-containing powder superalloy. Care must be exercised in assigning the hafnium L series peaks which extend from approximately 8 keV to 10.5 keV because they are overlapped by the copper K peaks and platinum L peaks which are due to excitation of the specimen support grid and cooling ring (see Introduction).

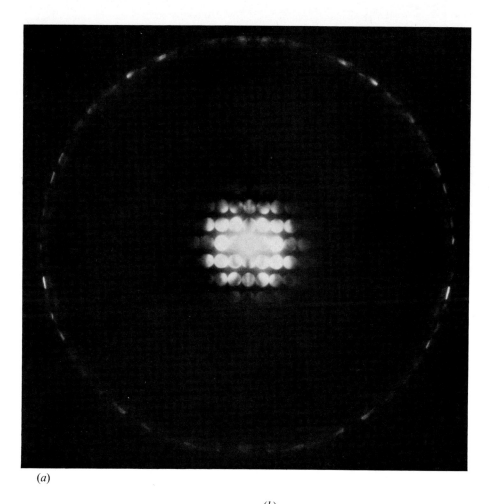

(a)

(b)

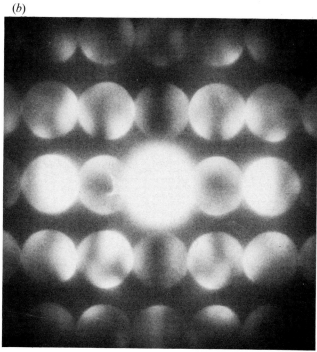

Figure 22.2 [101] hafnia zone axis pattern. (*a*) Short camera length. The symmetry of the zero-layer discs is not easy to see but the FOLZ ring reveals a single mirror line. (*b*) Long camera length. The most obvious features are the two sets of somewhat imperfect alternating dark bars in the vertical and horizontal rows of reflections. Since the low camera length micrograph revealed a single mirror there is a 2_1 screw axis perpendicular to it (horizontal) together with a vertical glide plane. In short, this single imperfect pattern fixes the space group of the crystal. Patterns recorded at 120 kV.

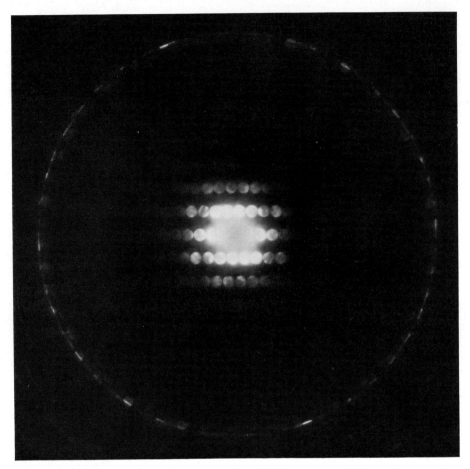

Figure 22.3 [010] hafnia zone axis pattern. The 2-fold symmetry of this pattern may be deduced from the FOLZ ring. Pattern recorded at 120 kV.

Phase name: silicon carbide, α-form
Lattice type: hexagonal
Point group: *6mmm*
Space group: *P6₃/mmc*
Lattice parameters: $a = 0.3076$ nm. $c = 0.5048$ nm,
$c/a = 1.6409$, 2H type [1]

SiC

FORMULA AND COMPOSITIONAL VARIABILITY

There are two forms of this material, α and β. The α form exists in a large number of polytypes (~20) with various stacking sequences. The β form is cubic. The SiC discovered on 316 steel CERs was the α polytype that has the wurtzite structure, but it was quite heavily faulted. The particles were presumably the result of mechanical polishing during specimen preparation.

OCCURRENCE

The phase was found on 316 CERs as small faulted flakes. The sizes ranged from fractions of a micron to several microns in length.

REFERENCE

1 Donnay J D H and Ondik H N (ed) 1973 *Crystal Data Determinative Tables, Vol 2—Inorganic Compounds* (US Dep. Comm., NBS and JCPDS)

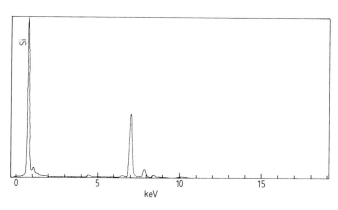

Figure 23.1 EDX spectrum from a silicon carbide chip discovered on a CER of a sample of 316 steel.

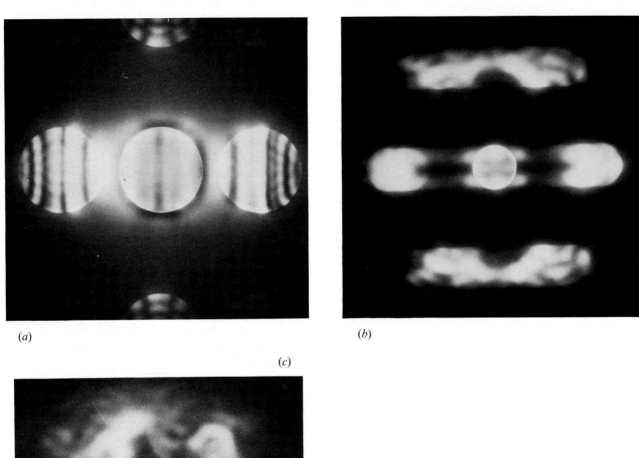

(a)

(b)

(c)

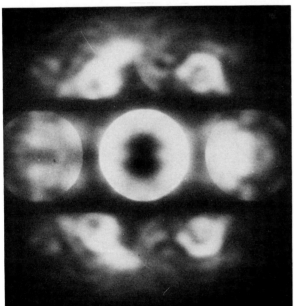

Figure 23.2 (*a*). ⟨10$\bar{1}$0⟩ SiC CBED pattern. It was not at all obvious that the SiC flakes were faulted at this zone axis. (*b*) and (*c*) are two ⟨11$\bar{2}$0⟩ patterns recorded with two different C2 apertures. Both show streaking (along [0001]) that is characteristic of heavy faulting parallel to the beam direction.

Appendices

Appendix 1

Glossary of abbreviations

ASM	American Society for Metals
bct	body-centred tetragonal
CB(E)D	convergent beam (electron) diffraction
CER	carbon extraction replica
C2	second condenser (lens or aperture)
EDX	energy dispersive x-ray (analysis)
fcc	face-centred cubic
FOLZ	first-order Laue zone
hcp	hexagonal close-packed
HOLZ	higher-order Laue zone
JCPDS	Joint Committee on Powder Diffraction Studies
SAD	selected area diffraction
SOLZ	second-order Laue zone
TMS-AIME	Transactions of the Metallurgical Society of the American Institute of Mining Engineers
ZAP	zone axis pattern

Appendix 2

Alloys used in this study

Table A2.1 Phases identified in each of the alloys studied in generating the material reproduced in this book.

AISI 316 stainless steel (Fe–Cr based)	Inconel 718 (Ni–Fe based)	Incoloy 901 (Ni–Fe based)	Astroloy (Ni based)
γ (austenite matrix)	γ (austenite matrix)	γ (austenite matrix)	γ (austenite matrix)
		γ'	γ'
	γ''		
δ (ferrite)	δ (orthorhombic)		
	α (ferrite)		
σ	σ		
Laves	Laves	Laves	
χ			
$M_{23}X_6$			$M_{23}X_6$
M_6X		M_6X	
MnS	MX	MX	MX
M_7X_3	M_7X_3		
M_2X			
M_3B_2	M_3B_2	M_3B_2	M_3B_2
M_2B			
	$Ti_4C_2S_2$		$Ti_4C_2S_2$
	MM$'$P		
TiO_2 (rutile and anatase)			
α-alumina			
AlN			
spinel			spinel
			ZrO_2 (HfO_2)
SiC			

Table A2.2 Compositions of the alloys in weight per cent.

	Ni	Cr	Co	Mo	Nb	Fe	Mn	Al	Ti	C	B	S	Si	P
Inconel 718	53.5	18.6	—	3.1	5.0	18.5	—	0.4	0.9	0.04	0.005	—	—	—
Incoloy 901	42.5	12.5	—	5.7	—	36.0	—	0.2	2.8	0.05	0.015	—	—	—
Astroloy	55.2	15.0	17.0	5.2	—	—	—	4.0	3.5	0.03	0.03	—	—	—
AISI 316	10–14	16–18	—	2–3	—	$\geqslant 62$	$\leqslant 2.0$	—	—	< 0.08	$\leqslant 0.01$	$\leqslant 0.05$	$\leqslant 1.0$	$\leqslant 0.03$

Appendix 3

Bracket conventions for Miller indices

Listed below are the standard conventions for Miller indices [1]:

(*hkl*)	indicates a particular set of crystal planes.
{*hkl*}	indicates a family of crystallographically equivalent planes.
[*uvw*]	indicates a particular crystallographic direction.
⟨*uvw*⟩	indicates a family of crystallographically equivalent directions.
hkl	indicates the reflection from a set of parallel planes; coordinates of a reciprocal lattice point.

Although we have adhered to these conventions in the text, we have dropped the brackets from the figures to save space. Where the figure contains indices referring to reflections and zone axes, as in figure 19.2, the zone axes have been written in square brackets. Reflections are referred to by indices without brackets in both the text and the figures.

Reference

1 Hahn T (ed) 1983 *International Tables for Crystallography* (Dordrecht: Reidel)

Appendix 4

Some aspects of crystal geometry

In a triclinic system, the dot product of two lattice vectors r_1 and r_2 may be written as

$$r_1 \cdot r_2 = (u_1 a + v_1 b + w_1 c) \cdot (u_2 a + v_2 b + w_2 c)$$
$$= u_1 u_2 a^2 + v_1 v_2 b^2 + w_1 w_2 c^2 + (v_1 w_2 + v_2 w_1)bc \cos \alpha$$
$$+ (w_1 u_2 + w_2 u_1)ca \cos \beta + (u_1 v_2 + u_2 v_1)ab \cos \gamma$$

where a, b, c are the magnitudes of the primitive translation vectors (lattice parameters) and α, β, γ are the interaxial angles. The included angle, ϕ, between two real space vectors r_1 and r_2 is given by

$$\cos \phi = \frac{r_1 \cdot r_2}{|r_1||r_2|}.$$

Here, $|r|^2 = r \cdot r$, and is given by

$$|r|^2 = u^2 a^2 + v^2 b^2 + w^2 c^2 + 2\,vwbc \cos \alpha$$
$$+ 2wuca \cos \beta + 2\,uvab \cos \gamma.$$

Similar formulae apply to reciprocal lattice vectors g_1, g_2, where $g = ha^* + kb^* + lc^*$, provided that a^*, b^*, c^*, and also α^*, β^*, γ^* are substituted in the above formulae.

Useful formulae [1] relating real and reciprocal parameters are listed below:

$$a^* = (bc \sin \alpha)/V \qquad b^* = (ac \sin \beta)/V \qquad c^* = (ab \sin \gamma)/V$$

with V, the volume of the unit cell, given by

$$V = abc \,(1 - \cos^2 \alpha - \cos^2 \beta - \cos^2 \gamma + 2 \cos \alpha \cos \beta \cos \gamma)^{1/2}.$$

$$\cos \alpha^* = (\cos \beta \cos \gamma - \cos \alpha)/\sin \beta \sin \gamma \qquad \cos \beta^* = (\cos \gamma \cos \alpha - \cos \beta)/\sin \gamma \sin \alpha$$

$$\cos \gamma^* = (\cos \alpha \cos \beta - \cos \gamma)/\sin \alpha \sin \beta$$

The included angle, θ, between a real vector r and a reciprocal vector g is given by:

$$\cos \theta = \frac{g \cdot r}{|g||r|} = \frac{(hu + kv + lw)}{|g||r|}$$

where $|g|$ and $|r|$ are calculated from the expressions above. The spacing of a set of reciprocal lattice planes normal to a zone axis U is simply $|U|^{-1}$ for a primitive lattice. In the case of crystal classes of higher symmetry than triclinic we obtain the results given in table A4.1.

For the hexagonal and rhombohedral crystal systems the above definition of the reciprocal lattice is not followed, and a special reciprocal lattice with base vectors in the basal plane, spaced by 120°, is used [2]. This special lattice proves to be more natural and useful than that formed by the general definition. It allows 4-index Miller–Bravais indices to be used to maximum advantage. The relationship between Miller indices and Miller–Bravais indices for the rhombohedral system is given below.

Miller indices		Miller–Bravais indices
$(h_R\, k_R\, l_R)$		$(h\, k\, i\, l)$
		$i = -(h+k)$

where $h_R = (2h+k+l)$ or $h = h_R - k_R$

$k_R = (-h+k+l)$ or $k = k_R - l_R$

$i = l_R - h_R$

$l_R = (-h-2k+l)$ or $l = h_R + k_R + l_R$.

Note also that the direction $[u_R\, v_R\, w_R]$ in Miller indices becomes $[u\, v\, t\, w]$ in Miller–Bravais indices.

The expression for $|U|^{-1}$ listed in column 3 of the table gives the spacing (H_{uvw}) of the reciprocal lattice planes normal to a $[uvw]$ zone axis, i.e.

$$H_{uvw} = |U|^{-1} \quad .$$

For non-primitive lattices we have $H_{uvw} = p|U|^{-1}$, where p is an integer. For a C lattice (monoclinic or orthorhombic), the conditions on p are:

$p = 2$ if u and v are odd and w even⎫ u, v, w are asssumed to be irreducible
$p = 1$ for all other u, v, w ⎭ integers throughout.

The equivalent conditions for A and B lattices follow by permutation of u, v and w. An F centred lattice (orthorhombic or cubic) includes A, B and C lattices and the conditions are:

$$p = 1 \text{ if } u+v+w = \text{odd}$$
$$p = 2 \text{ if } u+v+w = \text{even}.$$

For an I lattice (orthorhombic, tetragonal or cubic), the conditions on p are:

$$p = 2 \text{ if } u, v, w \text{ are all odd}$$
$$p = 1 \text{ otherwise}.$$

The value of H may be derived from ZAPS containing sharply defined HOLZ rings. The relationship between the radius of the FOLZ ring, G, and H may be written, to a good approximation, as:

$$G = \sqrt{(2\,K\,H)}$$

where K is the reciprocal of the electron wavelength (λ). The value of G should be corrected to take into account distortions produced in the diffraction plane by the microscope lenses [3].

References

1 Hahn T (ed) 1983 *International Tables for Crystallography* (Dordrecht: Reidel)
2 Okamoto P R and Thomas G 1968 *Phys. Stat. Sol.* **25** 81
3 Steeds J W 1981 Microanalysis by Convergent Beam Electron Diffraction, in *Microanalysis with High Spatial Resolution* (Proceedings of a conference sponsored by the Metals Society and the Institute of Physics, Manchester 1981) p 210

Table A4.1 Scalar products for general pairs of reciprocal ($g_1 \cdot g_2$) and real ($r_1 \cdot r_2$) lattice vectors, and the expression for the interplanar separation along a given zone axis.

Crystal class	$g_1 \cdot g_2$	$r_1 \cdot r_2$	$|U|^{-1}$
monoclinic	$\dfrac{h_1h_2}{a^2\sin^2\beta} + \dfrac{k_1k_2}{b^2} + \dfrac{l_1l_2}{c^2\sin^2\beta} - \dfrac{h_1l_2+l_2h_1}{ac\sin^2\beta}\cos\beta$	$a^2u_1u_2 + b^2v_1v_2 + c^2w_1w_2 + ac(w_1u_2+w_2u_1)\cos\beta$	$(u^2a^2 + v^2b^2 + w^2c^2 + 2uwac\cos\beta)^{-1/2}$
orthorhombic	$\dfrac{h_1h_2}{a^2} + \dfrac{k_1k_2}{b^2} + \dfrac{l_1l_2}{c^2}$	$a^2u_1u_2 + b^2v_1v_2 + c^2w_1w_2$	$(u^2a^2 + v^2b^2 + w^2c^2)^{-1/2}$
hexagonal and rhombohedral†	$\dfrac{2}{3a^2}(h_1h_2 + k_1k_2 + i_1i_2 + \lambda^{-2}l_1l_2)$	$\dfrac{3a^2}{2}(u_1u_2 + v_1v_2 + t_1t_2 + \lambda^2w_1w_2)$	$\left(\dfrac{3a^2}{2}(u^2+v^2+t^2+\lambda w^2)\right)^{-1/2}$
tetragonal	$\dfrac{h_1h_2+k_1k_2}{a^2} + \dfrac{l_1l_2}{c^2}$	$a^2(u_1u_2+v_1v_2) + c^2w_1w_2$	$[a^2(u^2+v^2)+c^2w^2]^{-1/2}$
cubic	$(h_1h_2+k_1k_2+l_1l_2)/a^2$	$a^2(u_1u_2+v_1v_2+w_1w_2)$	$a^{-1}(u^2+v^2+w^2)^{-1/2}$

† where $\lambda^2 = \frac{2}{3}(c/a)^2$.

Appendix 5

k factor calibration curve

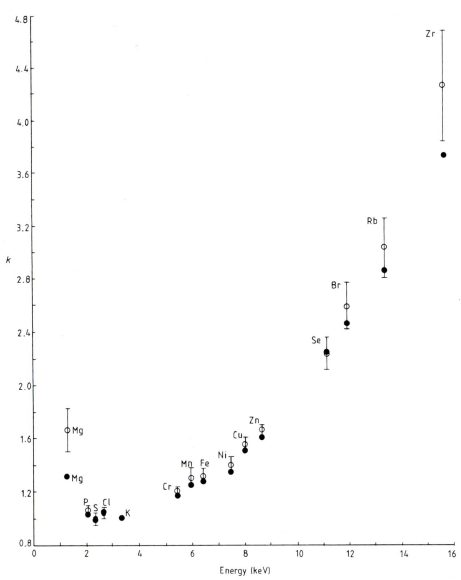

Figure A5.1 Comparison of experimental (○) and theoretical (●) *k* factors for the EDX system employed in the recording of the spectra presented in this work. A further explanation of this calibration curve can be found in 'Determination of Cliff–Lorimer *k* factors by analysis of crystallised microdroplets' (see reference [16] of Introduction).